Technology for Technology Education

Edited by
Robert McCormick, Charles Newey
and John Sparkes
at the Open University

Addison-Wesley Publishing Company

Wokingham, England • Reading, Massachusetts • Menlo Park, California
New York • Don Mills, Ontario • Amsterdam • Bonn • Sydney • Singapore
Tokyo • Madrid • San Juan • Milan • Paris • Mexico City • Seoul • Taipei

in association with

The Open
University

Cover designed by Designers & Partners of Oxford and printed by The Riverside Printing Co. (Reading) Ltd.
Typeset by Columns Design and Production Services Ltd, Reading, UK.
Printed in Great Britain by T.J. Press, Padstow, Cornwall.

First printed 1992.

This reader and its companion volume, *Teaching and Learning Technology*, are part of the course material for an Open University M.A. in Education module, E823 *Technology Education*. Open University students have access to other material as part of this module and hence the contents of this reader are necessarily a limited selection.

Anyone wishing to obtain more details about the module and the M.A. programme should contact The Central Enquiries Office, The Open University, Walton Hall, Milton Keynes MK7 6AA.

The views expressed in the articles in this collection are not necessarily those of the team responsible for developing the module, nor of the Open University.

British Library Cataloguing in Publication Data
A catalogue record for this book is available from the British Library.

ISBN 0 201 63168 7

Technology for Caerleon
Library
Technology Education

This book is due for return on or before the last date shown below.

Acknowledgements

The publisher has made every attempt to obtain permission to reproduce material in this book. Copyright holders of material which has not been acknowledged should contact the publisher. Grateful acknowledgement is made to the following sources for permission to reproduce material:

Text

Naughton, J. What is 'Technology?' In The Open University (1988) *T102 Living with Technology: Unit 1 Introduction*; Engineering design, (1984) *T392, Block 1*; Where are designers placed? (1983) *T263 Unit 1*; Control (1989) *T394, Control Engineering. Control Strategies, Unit 1*; A systems approach to food provision. (1989) *T274 Block 1*; Design project planning: case study of a litter bin. Harrison, M.E. and Sparkes, J. (eds) (1984) *T392, Block 3*; Walker, D. (forthcoming) The Fisher–Miller skeleton. Edited version of case study in OU Pack P791 Managing Design *E626 Design & Technology in the Primary Curriculum*, Resource Material: Palin, M. Food production: bread. Modified version of The Open University (1987) *T274, Unit 9*; British Home Stores and its point-of-sale system. *T102 Block 2*: Milton Keynes: Open University. Layton, D. (1991) Science education and praxis. In *Studies in Science Education* **19** pp. 43–56 © Studies in Education Ltd. Klemm, F. (1959) *A History of Western Technology*. London: George Allen & Unwin now Unwin Hyman of Harper-Collins Publishers Ltd. Cardwell, D.S.L. (1972) *Technology, Science and History*. London: Heinemann © D.S.L. Cardwell. Herschel, J.F.W. (1830) *Preliminary Discourse on the Study of Natural Philosophy* London: Longman. Rankine, W.J.M. (1972) *A Manual of Applied Mechanics*. London: Griffiin. Barnes, B. (1982) The science–technolgy relationship: a model and a query. In *Social Studies of Science* **12**, pp. 166–72 Reprinted with permission © 1982, Sage Publications Ltd. Archer, B. The Three Rs. In Cross, A. and McCormick, R. (eds) (1986) *Technology in Schools*. Design Studies 1 (1) pp. 17–20. Reproduced by permission of the publishers Butterworth-Heinemann Ltd. Lowe, I. Who's paying the driver. In Gale, F. and Lowe, I. (eds) *Changing Australia*. Irvine, S. No growth in a finite world. In *New Statesman & Society*, 23 November 1990 © New

Statesman & Society. MIT (1972) *Limits to Growth*. Wajcman, J. (1991) Feminist critiques of science and technology. In *Feminism Confronts Technology*. Polity Press. Barnes, B. and Edge, D. (eds) (1982) *Science in Context*. Milton Keynes: Open University Press. Cowan, R.S. (1970) From Virginia Dare to Virginia Slims: Women and Technology In *Technology and Culture*, 20 (1), pp. 51–63, Chicago: The University of Chicago Press. Cooley, M. (1991) *Architect or Bee?: The Price of Technology*. London: The Hogarth Press. Marcuse, H. (1968) *Negations*. London: Allen Lane. *Design*, 380 (August 1980). Roy, R. and Potter, S. *et al.* (1990) *Design and the economy*; Roy, R. and Potter, S. *et al.* (1990) Radical and incremental innovation: a tale of three trains. In *Design and the Economy*. London: The Design Council © The Design Council. British Standards Institution (1989) *Guide to Managing Product Design*. Persig, R.M. (1971) *Zen and the Art of Motorcycle Maintenance*. London: Jonathan Cape © R.M. Zen. Guterl, F. (1988) The compact disc. *IEEE Spectrum*, 25 (11) pp. 102–8. portions reprinted with permission © 1988 IEEE. Mayntz, R. and Hughes, T. (eds) The Development of Large Technical Systems. Max-Planck-Institut für Gesellschaftsforschung pp. 217–44 © Campus Verlag GmbH. Kay, W. (1985) How the Conran magic wand could transform BHS. *The Times*. Smith, P. Retail managers need a flair for making shopping a pleasing social activity today. *The Times*. Extracts from BS 7000 1989 are reproduced with the permission of BSI. Complete copies of the Standard can be obtained by post from BSI Publications, Linford Wood, Milton Keynes MK14 6LE.

Figures

Menai and Britannia bridges: The Hulton-Deutsch Collection. Post Bridge, Dartmoor; Royal Albert bridge, Saltash: The British Tourist Authority. Abbeywood Road Viaduct: Howard Humphreys & Partners Ltd Consulting Engineers. Dornoch Firth bridge – construction: Christiani and Nielsen. Dee Bridge railway accident: *Illustrated London News*, June 12 1847, p. 380. Taf Fechan bridge: Rendell Palmer & Tritton. Reproduced with permission STP Photography, Cardiff © STP Photography. Ponte de l'Alcantara e l'Alcazar: The Mansell Collection. Ironbridge, Coalbrookdale: The Science Museum. Thrust of Arches (Coulomb): Hyman, J. (1972) *Coulomb's Memoir on Statics*: Cambridge University Press. Dartford bridge: Dartford River Crossing Ltd. Pont de Luzancy: Freyssinet International (STUP) S.N.C. Box section: reproduced by permission of the Honorable the Speaker of the Legislative Assembly of Victoria, Australia. Forth Bridge: British Railways Board. Airfix model: Humbro Ltd. Model plane: Ripmax plc. Diagram of op-amp circuit; Ethanol model: Open University. Cut-away engine: Vauxhall Motors Ltd. Model aircraft in a wind tunnel: Crown Copyright, Defence Research Agency. Road map of London: Crown Copyright, Reed International Books. Under-ground map: London Regional Transport. Durham Cathedral © Erich Hartmann/John Hillelson Agency. APT and Intercity 225: Intercity.

Introduction:
In Search of Technology

R. McCormick, C. Newey and J. Sparkes

Technology as an aspect of human activity goes back much further than science. However, as David Layton argues in the second article in this Reader, it was not until relatively recently that technological activity has been recognized as something of equal status, and *different* from scientific activity. Science education has developed since the end of the last century, and there still remains much debate about how, and indeed what, science should be taught as part of general education. It is not surprising, therefore, that technology *education* is a newcomer in the school curriculum in most parts of the world. Equally it is not surprising that there is much debate about what should be taught. Some argue for teaching a problem-solving process with no specific content. Others focus almost solely on the content, particularly on 'hi-tech' aspects such as micro-electronics or computers.

The problem of defining what should be taught stems from the difficulty of adequately delineating what appears to be an enormous range of human activity. This range includes differences of scale from space programmes to a single person operating an engineering workshop. But it also includes such different kinds of activity as food production, electronics, medical technology, and mining. What is it that binds these all together under the label 'technology'? This Reader is an attempt to explore the links among the diverse activities, and to answer this question.

Our answer to the question of what constitutes technology is dealt with in all three parts of the Reader. The articles in Part 1 present some general ideas, by way of discussions of definitions of technology, and the distinctions between science, design and technology. One crucial difference between science and technology, highlighted by John Sparkes in Article 1.3, is the importance of value judgements in technological activity. It is with this in mind that Part 1 also explores value issues relating to gender, economic growth, and the attendant impact on the environment and resources.

The choice of articles in Part 2 of the Reader elaborates some of the features of technology that allow us to give the diverse activities a single label. Such things as modelling, design, planning and control are common to much technological activity, as is quality which, with the introduction of national standards, has taken on a new meaning within technological activity in industry. It is a meaning that has wide application and, along with quality assurance, constitutes a way of thinking that is relevant to all technological activity including that found in schools.

We believe that it is unhelpful to draw the boundaries of technology too tightly, and so in Part 3 the articles illustrate a range of specific technologies. Not only does the selection show some of the diversity of the activity, but also its complexity. This offers an opportunity to test the generalizations about the nature of technology found in the first two parts against the specifics of technology in the world around. We have tried to include activities that have some relevance to schools, but the selection is uncompromisingly a view of 'real-world' technology. Our profound belief is that, whatever reasons teachers have for teaching technology in schools, they must do so from some basis of an understanding of technology in the world around. In the same way as it would be inconceivable to teach science unrelated to how it is understood and practised in the world outside school, so too with technology. That poses significant problems, problems that are the province of the companion reader, *Teaching and Learning Technology*.

Although 'technology' as a curriculum subject has a recent pedigree in Britain and elsewhere, the nineteenth century also saw calls for technical work in schools for many of the same reasons we hear enunciated today. As we noted, the problem for schools in this 'technological age' is to be able to represent such things as nanotechnology and genetic engineering within the resources of schools and their teachers. Inevitably schools cannot hope to deal with such exotic activities, and must deal with that which is basic and enduring in technology. We hope this Reader helps to prepare teachers and those who support them to deal with one of the most profound influences on our society. The search for technology is therefore for us all, and our selection of articles is only a start, but a start which we hope will encourage you to join the search.

Some of the articles reprinted in this book have been edited. The use of square brackets and ellipses in these articles indicates that material has been added or deleted by the editors.

Contents

PART 1

General Ideas on Technology

1.1

What is 'Technology'?

J. Naughton

One of the things you will find is that academics seem to be obsessed with definitions, that is, with the meanings of words.

You will also discover as time goes on that there is no absolute standard for judging whether a given definition is right or wrong. Even referring to a dictionary will not solve the problem, because dictionaries *follow* rather than *lead* what people say. By that I mean that dictionaries (or, more accurately, their writers) do not *create* words, or dictate their meaning: they simply impose some order on the way in which we use words. This is why dictionaries always lag some distance behind the times, and why they have to be updated: new words are continually being created and existing ones modified to meet changing circumstances.

For example, the term 'satellite' originally meant 'an attendant upon a person of importance, forming part of his retinue and employed to execute his orders'. In other words it meant a kind of servant. As the science of astronomy developed, this idea of a satellite as a dependent person of low importance was applied to give the word a new meaning, namely 'a small or secondary planet which revolves around a larger one'. Nowadays, the word has been modified still further to mean an artificial object orbiting the earth (e.g. a 'tele-communications satellite').

This evolution is an on-going process. Even as I write, for example, I hear that 'satellite' is being used (incorrectly in my view) to refer to a certain kind of dish aerial used to receive television signals from orbiting satellites. Another example: the word 'transistor' was originally the name given to a small electronic component used to replace valves in electrical circuits. But it rapidly came to be used as the name for portable *radios* made with these components. Much the same has happened with the term 'video', which is now commonly used to refer both to video *tapes* and video *recorders*.

The purpose of this [article] is simply to identify two definitions of 'technology' and to argue for the adoption of one.

Broadly speaking, one of these definitions regards technology as *things*, whereas the other defines it in terms of an *activity* involving groups of people.

3

■ Technology as 'things'

Equating technology with machinery (sometimes spoken of as 'hardware') is very common. For example, I remember an advertisement for an expensive hi-fi system. Over a photograph of this magnificent piece of equipment was the caption: 'Isn't Technology Beautiful?'

This definition of technology as machinery clearly has its uses, and not just in advertisements. For example, in Block 1 of *Living with Technology* (Open University, 1988), the author talks about the 'technology of the home', and refers to a house as a piece of technology, 'a machine for living in', as a famous architect (Le Corbusier) once put it.

The equation of technology with machinery is clearly valid in the sense that it represents common usage, but it also has severe limitations. Consider, for example, the American moon programme of the 1960s (the Apollo Project, as it was named). The success of the programme was hailed as a 'major feat of modern technology'. This claim makes little sense if technology is defined solely in terms of machinery. For although the programme made use of sophisticated machines, the machinery by itself was not sufficient to account for the achievement of putting men on the moon and returning them safely to earth. Clearly, something more than machinery was involved.

What were the extra ingredients? Well, first of all there was a *goal*. This was set by President John F. Kennedy as getting a man onto the moon by the end of the 1960s. This goal was then broken down into a series of *practical tasks* – building rockets, sending astronauts into orbit, designing and testing lunar vehicles, etc. Secondly, there were *people*, namely scientists, engineers, technicians and computer experts, who were often very specialized and highly skilled. This implies that a third ingredient was *knowledge* of certain kinds.

As individuals, however, none of these people (and there were more than 40 000 of them at the height of the project) could have achieved the task, no matter how individually skilled they were. So the fourth ingredient was some form of *social organization* to manage and direct the combined effort of all the people involved; in this case it was the managerial structure of the National Aeronautics and Space Administration (NASA).

How might we combine these ingredients to form a definition of technology that is wider than just machinery?

■ Technology as a human activity

Here is a first shot at a definition that meets the above requirements:

> 'Technology is the application of *scientific knowledge* to *practical tasks* by *organizations* that involve *people* and *machines*.'

In fact this is almost good enough, but not quite. A brief critical examination will show why. [. . .]

Looking at the definition, you will note that I have cheated a bit. Having originally identified just 'knowledge' as an essential ingredient of technology, I have now specified a *special* kind of knowledge, namely the kind called 'scientific'. This may be a reasonable way of tightening up the definition, but you should not take it on trust. So let me ask the question: 'Is technology necessarily the application of only one type of knowledge – the scientific kind?' But that only begs another question: what is scientific knowledge?

[. . .]

☐ **What is scientific knowledge?**

Imagine you are listening to a conversation between two 10-year-old children. One of them is trying to pump up his bicycle tyre – with some difficulty, it would seem. (There's probably something wrong with the valve.)

'Ouch,' he says. 'This pump is getting hot.'
'Pumps always do,' replies his companion. 'I've noticed that.'
'But why do they get hot?' asks the first child.
'Dunno,' says his friend. 'They just do.'

The second child clearly possesses knowledge of some kind about the behaviour of bicycle pumps under pressure. But is it scientific?

[. . .]

The answer is 'no', and the clue to why it is not scientific is contained in the friend's answer to the boy's question – as we shall see in a moment.

Now imagine that an older sister arrives on the scene and is asked why the pump gets hot. Fresh from a school physics class, she has the answer pat. 'Because when you pump you're really compressing the air in the pump, and the Universal Gas Law says that the temperature of a gas is related to its volume and pressure.'

The answer might not have enlightened the 10-year-olds, but from our point of view it is sufficient to indicate that the older girl possesses *scientific* knowledge about this particular matter. The clue which indicates this is that she explains a particular phenomenon (the heating of a bicycle pump) in terms of a *general*, theoretical 'law'. This states that for any gas, its volume multiplied by the pressure under which it is kept is a quantity which is proportional to its temperature.

This law was proposed a long time ago as an explanation for the behaviour of gases under certain conditions and, having been tested over the years, has been found to be reasonably accurate for low pressures. It is a *scientific* law because (1) it can be tested (e.g. by experiments), and (2) because it explains a wide variety of different phenomena in one general, abstract statement. It is not, in other words, just a law of bicycle pumps: it applies to motor-driven compressors, refrigerators and similar machinery as well.

I described the gas law as being 'theoretical'. By that I mean that it is expressed in terms of abstract *concepts* (pressure, volume, temperature) rather than in terms of concrete things. And if you were to ask *why* the law holds, a scientist would explain it in terms of another, more fundamental set of theoretical statements called the molecular theory of gases. And this theory, in its turn, can be explained by a still more fundamental one, the atomic theory of matter.

I mention all this not to try to impress you, but simply to highlight two important features of scientific knowledge. The *first* is its tendency to explain everyday events, problems or phenomena in abstract, theoretical terms. The *second* is the tendency, described above, for scientific theories to be linked to (supported by) other theories at deeper levels of abstraction (see Figure 1).

[. . .]

□ **Not all technological knowledge is scientific**

Having gone through all this rigmarole, let's return to our original question: *Is technology just the application of scientific knowledge?*

The answer is *no*, because there are many examples of activities that are clearly 'technological' and yet involve types of knowledge *other than* the scientific kind.

For example, the construction of Durham Cathedral in the 11th and 12th centuries was a great technological achievement. The men who built it had to solve a very difficult practical problem, namely that of constructing a high, wide church with a stone ceiling or 'vault'. The problem arose in the first place

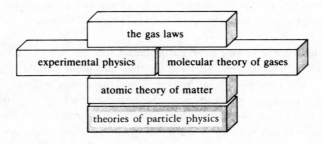

Figure 1 Scientific theories depend on other scientific theories.

because the physical properties of stone mean that it cannot be used over wide spans. The builders of the time had learnt about this deficiency by bitter experience (the collapse of earlier attempts at wide spans), but they did not know *why* stone had this property. They lacked, in other words, the scientific knowledge about the internal structure of materials that is provided by the modern specialism known as materials science.

Not only did the builders of Durham Cathedral lack a scientific explanation of their problem, but they were also able to solve it without recourse to science. What they did was to divide the vault into small areas of 'shell' by means of stone ribs, that is, arches which crossed the church both traversely and diagonally, as shown in Figure 2 (overleaf). By doing this they simultaneously made the roof stronger and lighter.

However, this brilliant technical solution in turn gave rise to another serious structural problem, because the forces resulting from the weight of the vault could not simply be supported by the piers and walls of the building. This was because these forces exerted an *outward* rather than a downward vertical thrust, thereby threatening to push the walls apart. Nowadays this problem could be routinely analysed using standard techniques of the science of engineering mechanics. The Durham builders knew nothing of this; nevertheless, they solved the problem by supporting the piers of the building using what are called 'flying buttresses' (see Figure 2 again).

This is a simple example, but it illustrates the point that technology need not necessarily involve the application of formal *scientific* knowledge. The cathedral builders in fact applied another kind of knowledge, namely the 'craft' knowledge they picked up from experience and passed on through successive generations of master masons.

Talking about medieval cathedrals may seem to you to be going too far back into the past, so let's consider a more modern example. Take a typical technological problem – the design of a motor car. Clearly there is a good deal of scientific knowledge involved: the science of aerodynamics guides the way the shape of the vehicle is designed to reduce wind resistance; the theory of mechanics and fluids helps engineers to design the combustion chambers in which the fuel is burned; the same scientific theories guide the design of suspension elements such as shock absorbers; the sciences of chemistry and materials guide the selection of particular rubber compounds for the tyres. And so on.

But even though a lot of scientific knowledge goes into the design of a car, the task cannot be accomplished without the other kinds of knowledge too. For example, one of the most important and distinctive features of a car is how it 'feels' to the driver: is it responsive, lively, stable? Or sluggish and heavy? Does it look elegant, or ugly, or functional? Does the driving position 'feel' right to you? There are few *scientific* theories that designers can draw on to help them here: they have to look to other sources of knowledge such as experience, design and craft knowledge and to their own feelings for particular configurations. Science, in other words, though vital to the technologist, is not enough.

stone, shell-like
infilling between ribs

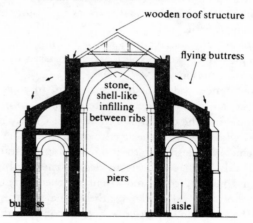

Figure 2 The structure of Durham Cathedral. Arrows show how thrusts (forces) from the ribs are carried partly by the flying buttresses.

Nor is 'knowledge about things', such as craft knowledge, the only thing the technologist needs to supplement science. He or she also needs knowledge about how to make things happen in organizations. Consider our previous example, the US Moon Programme (the Apollo Project). Here again the same point holds. For although much of the knowledge applied to the design of the machinery necessary to land on the moon *was* scientific, the managerial skills and knowledge necessary to manage the whole gigantic enterprise effectively were definitely not scientific. For there is no 'management science' in the sense that there is, say, a 'science' of physics. So you have to concede that, once again, forms of knowledge other than scientific were also involved in this particular technology.

[. . .]

This leads me to amend my definition in the following way: 'Technology is the application of scientific *and other knowledge* to practical tasks by organizations that involve people and machines.'

[. . .]

■ Some implications of the definition

I would like to make sure that you understand the implications of defining 'technology' in the way I have just done. I want to stand back from the definition for a moment to see its significance.

First of all, I have defined 'technology' as a *practical* activity. Its goal is to solve a problem, to make something happen. In that sense, technology is very different from science, because the goal of science is *understanding* not action.

[. . .]

Secondly, my definition says that technology involves applying not just scientific knowledge but also other types of knowledge. Although the truth of this is recognized by every practising engineer, it tends to alarm university teachers, because it suggests that technology involves not just the respectable theoretical knowledge that you get from studying science, but also the more uncertain practical knowledge you get from experience, craft, apprenticeship and other sources. [. . .] But because the knowledge in question is not theoretical it doesn't mean that it isn't important, or that we should somehow deny its existence in the practice of technology.

The third implication of my definition of 'technology' is that technology invariably involves people and organizations as well as machines. That means that it also *involves ways of doing things*. The whole thing is always a complex interaction between people and social structures on the one hand, and machines on the other. New technology, in this sense, does not just involve new machines: it also involves new ways of working, and perhaps new types of organization too.

■ Reference

The Open University (1988) *T102 Living with Technology*. Milton Keynes: The Open
University Press.

Science Education and Praxis: the Relationship of School Science to Practical Action

D. Layton

[In the original article from which this extract is taken, David Layton addresses the link between the kind of knowledge that is taught in school science lessons and practical action in the made world. He does this by first examining the relationship of science and technology through historical, philosophical and sociological perspectives, and it is this part of the original article that we reproduce here. (Reader 2 *Teaching and Learning Technology* contains an extract from a later part of the article.)]

[. . .]

■ The separation of science and technology

For Leonardo da Vinci, as for other Renaissance engineers (Gille, 1964), and for Francis Bacon some time later, knowledge of the natural world was the means by which the forces of nature might be controlled and harnessed to man's will. According to Bacon's aphorism, 'Human knowledge and human power meet in one' (Bacon, 1905, p. 259). Indeed, for him, the guarantee of the truth of knowledge was the extent to which it could be used for the relief of man's estate. 'For fruits and works are as it were sponsors and surities for the truth of philosophies', he declared (Bacon, 1905, p. 276), and 'truth and utility are here the very same thing'.

Whereas a recent interpretative study of Bacon and 'the maker's knowledge tradition' (Pérez-Ramos, 1988) raises doubts that some of his key notions, such as 'control over nature', had the same meaning for him as they do for us today, the Renaissance bonding of cognition and practical action is well documented. Gernot Böhme and colleagues from the Max-Planck Institut, Starnberg, cite Leonardo in support of the contemporary view that the purpose of natural knowledge was not only to discover the facts but also to construct 'arte-facts' according to the rules which delineated the realm of the natural world. 'If you were to ask me, "what do your rules bring about and of what use are they?" I should answer you that they prevent the inventors and investigators from promising themselves, and unto others, things that are impossible', Leonardo wrote (Böhme *et al.*, 1978, p. 223). With those experimentalists and operatives who shared this perception of the intimate association of knowing and doing, Leonardo and Bacon foresaw, in our terms, a simultaneous and unified scientific and technological revolution ahead.

It did not happen. Science and technology were to embark upon processes of development which followed separate, though inter-connecting, paths. The industrial revolution in Britain came a century after the scientific revolution and, with the possible exception of the chemical industry, its origins owed little to scientific knowledge (Russell, 1983, p. 99). Cognitive and institutional differentiation characterized science and technology in the period following the principate of Newton.

In the cognitive realm, the transformation of terrestrial and celestial mechanics by Galileo and Newton involved what Edwin Layton Jr. (1990) has called 'a heroic idealization of reality', the creation of a mathematical 'shadow world' of points which occupied no space, of material bodies unblemished by departures from perfect rigidity and glabreity and for which linear inertial motion (or rest) was the norm, of fluid media untroubled by turbulence and eddies, and of a space which was homogeneous and isotropic. In this abstracted world there was no intrusion of those scale effects which disconcerted engineers and with which Galileo had once wrestled (e.g. why large 'machines', built to the same geometrical proportions as smaller and operationally effective ones, often failed). The same laws of mechanics described events on both a terrestrial and celestial scale.

A notable feature of this indisputably potent transformation was the new goal of research to which it testified. Although scientists often began their investigations from some point of technological interest (e.g. the observation by mining engineers in the 16th century that 'suction pumps' could not raise water to a height greater than about 30 feet), their goal was not improved performance of the technological artefact, but an understanding of its principles leading to general theory (e.g. the subsequent work of Galileo's pupils, Torricelli and Viviani, together with that of Pascal, yielded a general theory of pneumatics). As Peter Weingart (1978, p. 265) has expressed it, 'The crucial point is that the search for causes transcends human artefacts and leads to the underlying principles of nature'. In these terms, science and technology became differentiated not by virtue of distinctive operating procedures nor even by conceptual frameworks, but primarily because of differences in goals. In the words of one distinguished American historian of technology, the divisions between them became ones 'between communities that value knowing and doing, respectively' (Layton, 1977, p. 209).

Of course, the communities interacted to varying degrees. The technologies of instrumentation, yielding clocks, thermometers, barometers, voltmeters, ammeters and the like, allowed greater precision and standardization of scientific data. Other technological products such as telescopes, microscopes, vacuum pumps and electric cells extended the range of sensory experience and assisted the reliable reproduction of phenomena. In turn, general scientific theory was redirected towards technological devices and problems, although not always with the success its proponents anticipated. Engineers were quick to point out the differences between, for example, the scientific mechanics of Galileo and Newton and the technical mechanics needed to ensure the effective operation of actual machines in practical situations. At the opening of the 18th century, Antoine Parent employed the newly available calculus and calculations consistent with Newtonian mechanics to show that the maximum useful effort from an undershot water-wheel was a mere 4/27 of the natural effort of the empowering water flow. Later in the century, from his practical knowledge of designing, constructing and repairing water-wheels, John Smeaton challenged this conclusion. By systematic and quantitative experimentation with a two-foot model wheel, using loads which ranged from zero to the limit of the wheel, Smeaton demonstrated that the maximum ratio was close to 1/3 for an undershot wheel and double this for an overshot wheel (Cardwell, 1972a, p. 79). In similar vein, Frederick the Great wrote to Voltaire in 1778 about the respective merits of 'theory' and 'practice'.

> 'The English have built ships with the most advantangeous section in Newton's opinion, but their admirals have assured me that these ships do not sail nearly so well as those built according to the rules of experience. I wanted to make a fountain in my garden. Euler calculated the output of the wheels which should have raised the water into the reservoir, from which it was to flow again through canals and again mount on high in the fountains at Sans Souci. My lifting-gear was carried out according to mathematical calculations but could not raise a drop of water to fifty paces from the reservoir. Vanity of vanities! Vanity of mathematics!' (*Klemm, 1959, p. 262*)

To a large degree, the communities of natural philosophers and practical men remained institutionally distinct throughout the 18th century, knowledge being transmitted in the former in terms of theory encoded and manipulated by means of abstract symbols and in the latter by personal demonstration and emulation. Indeed, battles over inconsistencies between 'theory' and 'practice' were to run on throughout much of the 19th century (e.g. Hunt, 1983; Bud and Roberts, 1984; Kline, 1987). In Smeaton, however, we can see the prototype of a 'new man', the professional engineer, who could mediate between the scientists, on the one hand, and the working mechanics, on the other. In addition to his studies on water-wheels, he used his method of systematic experimentation, over a four year period and involving more than 130 experiments, to bring about notable improvements in the efficiency of the Newcomen steam engine.

> 'His practice was to adjust the engine to good working order, and then after making a careful observation of its performance in that state, some one circumstance was altered, in quantity or proportion, and then the effect of the engine was tried under such change; all other circumstances except the one which was the object of the experiment, being kept as nearly as possible unchanged.' (*Cardwell, 1972a, p. 83 citing John Farey*)

Smeaton was a painstaking and pioneering exponent of the method of incremental and progressive improvements to existing technological artefacts. By analogy with Kuhn's 'normal' and 'revolutionary' versions of science, Smeaton's work can be described as 'normal technology' (Kuhn, 1962). Something more than systematic investigation was needed to go beyond optimization of design parameters, however, and to achieve radical innovation in design conception and realization.

■ The scientification of technology

One judgement on the conditions necessary for the progress of technology was provided early in the 19th century by J.F.W. Herschel in his *Preliminary Discourse on the Study of Natural Philosophy*. The practical arts remained, in his opinion, 'separated from . . . science by a wide gulf which can only be passed by a powerful spring. They form their language and their own conventions, which none but artists can understand. The whole tendency of empirical art, is to bury itself in technicalities, and to place its pride in particular short cuts and mysteries known only to adepts . . .' In contrast to the privately owned, artefact-specific and frequently tacit knowledge associated with much technological practice, science delighted in laying itself open to inquiry and in making the road to its conclusions 'broad and beaten'. In relation to technological artefacts, the collectivist aim of science was 'to strip away all mystery, to illuminate every dark recess, and to gain free access to all processes, with a view to improve them on rational principles' (Herschel, 1830, pp. 71–2).

Endorsement of this view came from William Whewell, addressing the third annual meeting of the British Association for the Advancement of Science at Cambridge in 1833, the context being an on-going deliberation about the extent to which invention and technological progress in the mechanical and chemical arts should have a place in the programmes of the new body. Eventually, an additional section, G, for mechanical *sciences* was created in 1836, but not before an ideology was in place, proclaiming the superiority of 'science, knowledge and theory' over 'practice'. 'Art has ever been the mother of Science', Whewell allowed, but science was 'a daughter of a far loftier and serener beauty' (Whewell, 1834, p. xxv).

Compensation to 'practical men' for their subordinate role in the science–technology relationship was offered by Whewell's first, and questionable, proposition that technology was the chief begetter of science. Additionally, there was the opportunity which the prospect of 'mechanical *sciences*' offered to upgrade their activity by its association with general theory and abstract symbolism. Incipient in the relationship was the basis for a later social contract whereby science would 'lay a succession of golden eggs, and . . . society should pay to understand how nature works in order to exploit the potentialities of nature' (Keller, 1984, p. 160). The resources of scientific theory could provide knowledge, principles and insights which were unlikely to become available from any Smeatonian, step-by-step, continuous development of a technological device. Access to them opened up new design and realization possibilities and hence the prospect of 'revolutionary' as opposed to 'normal' technological change. Theory was cast in the role of 'an heuristic of invention' (Böhme *et al.*, 1978, p. 236). While by no means the only source of novelty and artefactual discontinuity, science was portrayed as particularly potent, to an extent that encouraged a perception of technology as merely 'applied science'. Taken to the limit, an autonomous 'pure science' became less a resource to support developments in technology than an engine impelling technological change. As Lyon Playfair expressed it in a lecture following the Great Exhibition of 1851, 'the cultivators of abstract science, the searchers after truth for truth's sake are . . . the "horses" of the chariot of industry' (Cardwell, 1972b, p. 81).

Such a view, buttressed by plausible exemplars such as the electric telegraph and the synthetic dye industry, achieved wide credence, not least in educational circles. When in 1857, during his presidential address to the Chemical Society, Professor W.A. Miller, King's College London, acclaimed Perkin's discovery of mauve as 'a successful application of abstract science to an important practical purpose' he was doing no more than illustrate an already well burnished theme (Miller, 1857, p. 187). Of course, his claim was not strictly true, as Perkin's discovery had been accidental. More importantly, and typically, there was little indication that the act of application was anything other than unproblematical and routine. In reality, the task of scaling up from a laboratory bench experiment to the first multi-step, hazard-contained, industrial synthesis, yielding a product of quality and price acceptable to a substantial market, threw up formidable problems, not only scientific and technical, but economic, environmental and legal also (Travis, 1990).

The educational message systems of curriculum, pedagogy and assessment were used increasingly in the second half of the 19th century to define and promote 'pure science' as a dominant category detached from practical contexts. The social processes by which this acquired hegemony in educational institutions have been described elsewhere (Layton, 1973, 1975 and 1981; Bud and Roberts, 1984) and need not detain us here. The important point is that by the time economic and industrial influences, along with the professionalization of engineering, led to the introduction of technology into the university curriculum in Britain, 'abstract theory' had secured the high ground, as well as high status, while 'practice' was seen as demeaning and lacking in prestige.

This situation posed severe difficulties for those appointed to university chairs of engineering. If their course of study emphasized 'theory', this might be seen as an unwelcome encroachment on the territory of existing science departments; if the bias was towards 'practice', the danger was that they usurped the prevailing systems of pupilage and apprenticeship. The solution developed by W.J.M. Rankine, Regius Professor of Civil Engineering and Mechanics at Glasgow from 1855 until 1872, was to transcend the traditional categories of 'theory' and 'practice' by focusing on the nature of the interaction between them. To equate this with the application of science was, according to Rankine, to misrepresent it, unless it was understood that the process was an active one which often entailed the creative reworking of the science. For him, the 'application of these (scientific) principles to practice is an art in itself' (Rankine, 1857, p. 13).

Because science and technology had evolved with their own conceptual frameworks, discoveries in the former could not always be applied readily to the latter. The 'scientification' of technology by Rankine and others, such as Osborne Reynolds at Manchester and Fleeming Jenkin at Edinburgh, entailed the construction of new knowledge which functioned as an intermediary between abstract science and practical action. In this way, a problem in the micropolitics of the university curriculum was solved; at the same time 'engineering science' assumed a distinctive and autonomous character.

■ Technological knowledge

In drawing upon research in the history, philosophy and sociology of technology in order to elucidate further the nature of technology, it is noteworthy how recent most of this work is. The Society for the History of Technology was founded in 1958 and its journal *Technology and Culture*, pre-eminent in the field, has been published since 1959. A pioneering *Bibliography of the Philosophy of Technology* by Carl Mitcham and Robert Mackay appeared first as a special number of *Technology and Culture* in 1975 and volume one of *Research in Philosophy and Technology*, the official publication of the Society for Philosophy and Technology, was published in 1978. A more detailed account of the development of philosophy of technology is provided by Durbin (1989). While significant European (although not British to any extent) contributions to the history and philosophy of technology predate the predominantly North American developments noted above, the evidence supports the broad generalization that institutionalized concern about the nature of technology is a phenomenon of the second half of the 20th century. Furthermore, it is interesting to compare this with institutionalized concern about the nature of science in the first half of the century. *Isis*, the leading journal for the history of science, commenced publication in 1912. The first British university department for the history and philosophy of science, at University College London, was founded in 1923. Karl Popper's seminal *Logik der Forschung* was published in Vienna in 1934, and several journals such as *Annals of Science*, *Ambix* and *Notes and Records of the Royal Society* date from the mid-thirties also. J.B. Conant's influential (especially on science education) *On Understanding Science. An Historical Approach* was written in 1946. It is a reasonable speculation that, just as scholarly understandings of the nature of science have been used to influence the goals and practice of science education (Layton, 1990), so the results of research into the nature of technology will have a curriculum impact on technology education.

A case in point concerns the relationship between science and technology. Few, if any, of those who have studied this question would now subscribe to the hierarchical dependence model which portrays technology as subservient and merely involving the routine and menial application of scientific knowledge and technique. Meticulous historical studies of specific technological innovations in diverse fields confirm the untenability of this view. Edward Constant's (1980) account of the origins of the turbojet revolution, Thomas Hughes' (1983) prize-winning *Networks of Power: Electrification in Western Society*, Walter Vincenti's (1982) description of engineers' use of 'control-volume analysis' to illustrate 'a difference in thinking between engineering and physics', Hugh Aitken's (1985) research into the origins of radio and Ronald Kline's (1987) study of the development of the induction motor are just a few recent examples from an extensive field. With respect to science, technology is no longer judged to be subordinate; rather, the relationship is characterized by equality, symbiosis and interaction.

An elegant presentation of this thesis was provided by Edwin Layton (1987) in his Presidential Address to the Society for the History of Technology, entitled 'Through the looking glass, or news from Lake Mirror Image'. The title implies that science and technology are mirror-image twins, often indistinguishable to a casual observer who sees them, especially today, using similar equipment and apparently doing similar things. However, what they are about has different ends; both create knowledge, but 'technological knowledge is tailored to serve the needs of design. That in the basic sciences is shaped by the desire to construct the most general and comprehensive theory' (p. 605).

In a remarkable historiographical analysis of the language, methodology and predominant themes in articles published in *Technology and Culture* between 1959 and 1980, John M. Staudenmaier (1985) elaborated this thesis further. The relationship between science and technology had been the theme of two special issues of *Technology and Culture* in 1961 and 1976, and other articles had also addressed the topic. Staudenmaier drew attention to an important conclusion which seemed to be emerging from analysis of this research. Out of a growing consensus that 'practical usable criteria for making sharp neat distinctions between science and technology do not exist' (Mayr, 1976, p. 668) and that 'the categories of analysis, "science" and "technology", are not illuminating categories for understanding these activities' (Thackray, 1976, p. 645), a theme of greater potential had arisen, that of 'the characteristics of technological knowledge' (Staudenmaier, 1985, p. 85).

Technological knowledge is to be understood here as knowledge 'structured by the tension between the demands of functional design and the specific constraints of its ambience'. Design concepts cannot remain on the abstract level, but 'must be continually restructured by the demands of available materials, which are themselves governed by further constraints of cost and time pressures and the abilities of available personnel' (p. 104). The integration of 'the abstract universality of a design concept and the necessarily specific constraints of each ambience in which it operates' would seem to be the primary cognitive problem of technological knowledge (p. 111).

Staudenmaier identifies four characteristics of technological knowledge from his review of articles in *Technology and Culture*. These are designated 'scientific concepts'; 'problematic data'; 'engineering theory'; and 'technical skill' (pp. 103–120). [See Figure 1] Two points about the first of these are of particular significance for science education. In their investigations into the role of science in technological developments, numerous authors provided evidence that 'before scientific concepts can contribute to technological knowledge they must be appropriated and restructured according to the specific demands of the design problem at hand' (p. 104). They do not remain in their original form. Furthermore, viewed from the perspective of technological knowledge as unique and irreducible, the intellectual role of science in the science–technology relationship is no longer that of the superior partner. Indeed, it could be argued that 'pure science is but a servant to technology, a charwoman serving technological progress' (Skolimowski, 1966, p. 373).

'Problematic data', the second characteristic, refers to those areas of ignorance with which technologists are frequently confronted. As Rankine had explained, it was rarely possible to wait for the progress of science; immediate

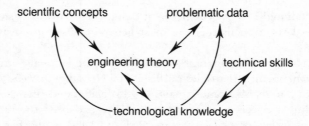

Figure 1   A model of the nature of technological knowledge (after Staudenmaier, 1985).

action was expected from the technologist and if existing data did not permit an exact solution to a problem, then the best approximation had to be acted on. 'A prompt and sound judgement in cases of this kind is one of the characteristics of a *practical man*' (Rankine, 1872, p. 10). The search for problematic data can arise in a variety of contexts such as the development of a new technology, the catastrophic failure of an established technology and the activities of regulatory and health and safety agencies. Staudenmaier's conclusion that 'No technology is ever completely understood, even after it has been introduced into normal practice' (p. 107) is endorsed by Brian Wynne's study of technological accidents, subtly entitled 'Unruly Technology'. Using his evidence to comment on the social nature of 'normal' technology, Wynne argues that 'the operating rules of technologies are an *ad hoc* brew of informal modes accommodating imprecise general principles to particular circumstances of implementation' (Wynne, 1988, p. 149). In contrast to science, where data are related to abstract and general theories, technological data are determined by and related to the specifics of technological practice.

Engineering theory is a primary characteristic of technological know-ledge, both scientific concepts and problematic data featuring as elements of it. From the way in which the term has been used by historians of technology, Staudenmaier defines it as 'a body of knowledge using experimental methods to construct a formal and mathematically structured intellectual system . . . (which) explains the behavioural characteristics of a particular class of artifact or artifact-related materials' (p. 108). This may sound like a definition of a scientific theory but, it is argued, it differs from this in both substance and style because its content and procedures are 'intellectually structured by the demands of technological praxis rather than by the more abstract demands of a scientific discipline' (p. 109). It would seem to be what Rankine envisaged when he claimed that his course of study would qualify a student 'to plan a structure or machine for a given purpose, without the necessity of copying some existing example, and to adapt his designs to which no existing example affords a parallel. It enabled him to compute the theoretical limit of the strength or stability of a structure, or the efficiency of a machine of a particular kind, – to ascertain how far an actual structure or machine fails to attain that limit, – to discover the causes of such shortcomings, – and to devise improvements for obviating such causes; and it enables him to judge how far an established practical rule is founded on reason, how far on mere custom, and how far on error' (Rankine, 1872, p. 9).

Accepting this, however, the fact remains that engineering theory is just that, a creation in the realms of thought and not a practical achievement in the made world. The articulation between theory and action still needs to be addressed and this necessitates consideration of the concept of technical skill.

One interpretation here is that the notion of engineering theory implies a division of labour between the theoretical expert who designs and plans, on the one hand, and the technically skilled operative who implements the design, on the other. A disjunction exists between knowing and doing. In other words, technical skill is nothing more than the application of that codified knowledge, in the form of formulae, rules and tables, which constitutes engineering theory.

A contrary view is that technical skill has to be learned experientially, that, following Nasmyth, 'The nature and properties of the materials must come in through the finger ends' and that 'no technological praxis is completely reducible to abstract theory' (Staudenmaier, 1985, p. 115). On this basis, there are technological judgements to be made which cannot be based on theoretical knowledge alone. One example is provided by the history of a project developing the German V-2 rocket. 'A lot of technical knowledge, common sense, and experience must be expected from the chief of such an organization . . . to determine the correct moment to freeze development, and to start production' (Dornberger, 1963, p. 400).

Taking a broad view of technological developments, including recent technological changes, Staudenmaier concludes that the work of historians of technology favours this second view of technical skill. A disjunction between knowing and doing, on which 'application' models depend, is not supported by the evidence, and the dominant interpretation is of technological knowledge as a unique and irreducible cognitive mode.

There is much in common here with what Donald Schön, writing about the nature of skilful professional practice, has termed 'knowledge-in-action', meaning by this 'a kind of knowing (which) is inherent in intelligent action' (Schön, 1983, p. 50). Like Schön, historians of technology reject as inadequate an epistemology of practice which is based on technical rationality, that is, 'the application of privileged knowledge to instrumental problems of practice' (Schön, 1987, p. xi). Emphasis on the tacit is more overt in Schön's account than in the historians' interpretation of technological knowledge and the latter perhaps too readily accepts in its references to science what Bruno Latour has called 'the two myths of universality and transparence, according to which science was not only the same everywhere, but also was the only form of knowledge completely explicit and articulated' (Latour, 1990, p. 96). The convergence of these different lines of enquiry on a distinct form of cognition associated with practical action is significant, however, not least for a science education which has been widely deemed to be laying the foundation upon which praxis ultimately rests.

■ Scientific knowledge and the everyday world

We turn now to another vein of evidence about the relationship between science education and practical action. Entry into it is provided by a reflection on the nature of the science–technology relationship which, as we have seen, historians of technology have relocated as a subtheme of a more comprehensive understanding of technological praxis as a form of knowledge.

In a critique of the hierarchical dependence model and advocacy of the preferred egalitarian, interactive model, the sociologist of science Barry Barnes (1982, p. 169) posed an interesting question. 'Why' he asked, 'should an interactive model of this kind not be used as a way of conceptualizing the relationship of science with other sub-cultures (than technology)? Why, for example, should the relationship between science and political sub-cultures, to the extent that there is such a relationship, not be conceptualized in this way, or the relationship between science and our everyday commonsense culture?' [See Figure 2]

[. . .] What Barnes is suggesting here is that the political contexts in which public policy is shaped on matters such as the standards to be applied to the use of a food additive, a new drug or a pesticide, interact with the

	'Bad old days'	Present
General image	S ↑ ↓ T hierarchical dependence	S ⟵⟶ T egalitarian, interactive
Main mediating agencies	words	people
Outcomes (a) for the development of knowledge	(a) predictable consequences. T deduces the implications of S and gives them physical representation. No feedback from T to S.	(a) no predictable consequences. T makes occasional creative use of S. S makes occasional creative use of T. Interaction.
(b) for the development of competence and technique.	(b) S may make free creative use of T as a resource in research.	(b) not a separate question. Interaction as above.
(c) for the evaluation of knowledge and competence.	(c) S evaluates discoveries in an unchanging context-independent way. T is evaluated according to its ability to infer the implications of S. Success in T is proper use of S: failure in T is incompetent use of S.	(c) S and T, both being inventive, both involve evaluation in terms of ends. No a priori reason why activity in T should not be evaluated by reference to ends relevant to agents in S, or vice versa.

Figure 2   Relationship between science (S) and technology (T) (after Barnes, 1982, p. 167).

science in significant ways. Edwin Levy terms the product 'mandated science', meaning by this 'the work of scientists and technologists in the context of bodies mandated to make recommendations or decisions of a policy or legal nature' (Levy, 1989, p. 41). Regulatory agencies, expert commissions, standard-setting organizations and law courts are examples of bodies likely to be involved.

Levy's response to Barnes' question is that it does indeed prove fruitful to conceptualize the relationship as interactive. (Scientific) 'data and papers must often undergo a kind of translation process when they enter the mandated area' because 'the standards of proof, nature of evidence, concepts of cause and effect, control of the situation' are not the same as in the context of a 'pure science' discipline. He contrasts the scientific paper which would be delivered by a scientist when addressing a conference of disciplinary peers with the substance and style of communication which the same scientist would need to use if appearing as an expert witness in a lawsuit (p. 43).

Levy's extension of his argument into the area of risk assessment need not be followed here, other than to note his rejection of the two-step model whereby 'science discloses; society disposes' i.e. *measurement of risk* is a scientific, probabilistic and objective activity while *acceptability of risk* involves personal and social value judgements. For him, both *measurement* and *acceptability* of risk are value-laden activities, and both involve science. While this is now an unexceptional and well-supported conclusion, its significance here is due to the light it throws on 'mandated science'. This is produced under 'messy, real-world conditions', far removed from the idealized and isolated systems of 'pure science' (p. 50). It is afflicted by uncertainties and fragilities with regard both to data and to interpretations which can be made; yet it is also required to serve as the basis for practical action. Its construction has much in common with that of technological knowledge as outlined in the previous section. Writing about the same phenomenon, which he terms 'citizen's science', J.R. Ravetz argues that 'all participants will need to cope with uncertainty and ignorance; to evaluate weak or ill-formed technical information; to distinguish (*not* separate) the more and less value-laden aspects of materials and methods; . . . and to devise appropriate political structures for consensus, decision, monitoring and enforcement' (Ravetz, 1985, p. 3).

[. . .]

■ References

Aitken, H.G.J. (1985) *Syntony and Spark. The Origins of Radio*. Princeton, NJ: Princeton University Press.

Bacon, F. (1905) *The Philosophical Works of Francis Bacon*. Reprinted from the texts and translations . . . of Ellis and Spedding (Robertson, J.M., ed.). London: G. Routledge and Sons.

Barnes, B. (1982) The science–technology relationship: a model and a query. *Social Studies of Science*, 12, 166–72.

Böhme, G., Van der Daele, W. and Krohn, W. (1978) The 'scientification' of technology. In Krohn, W., Layton Jr., E.T. and Weingart, P. (eds) *The Dynamics of Science and Technology*. Dordrecht, Holland: D. Reidel Publishing Company, 219–50.

Bud, R.F. and Roberts, G.K. (1984) *Science Versus Practice*. Manchester: Manchester University Press.

Cardwell, D.S.L. (1972a) *Technology, Science and History*. London: Heinemann.

Cardwell, D.S.L. (1972b) *The Organization of Science in England*, revised edn. London: Heinemann.

Constant, E.W. (1980) *The Origins of the Turbojet Revolution*. Baltimore: Johns Hopkins University Press.

Dornberger, W.R. (1963) The German V-2. *Technology and Culture*, 4, 393–409.

Durbin, P.T. (1989) History and philosophy of technology: tensions and complementarities. In Cutliffe, S.H. and Post, R.C. (eds) *Context. History and the History of Technology*. Bethlehem: Lehigh University Press; London: Associated University Press, 120–32.

Gille, B. (1964) *Les Ingénieurs de la Renaissance*. Paris: Hermann.

Herschel, J.F.W. (1830) *Preliminary Discourse on the Study of Natural Philosophy* (new edition). London: Longman.

Hughes, T.P. (1983) *Networks of Power. Electrification in Western Society 1880–1930*. Baltimore: Johns Hopkins University.

Hunt, B.J. (1983) 'Practice vs. Theory'. The British electrical debate, 1888–1891. *Isis*, 74, 341–55.

Keller, A. (1984) Has science created technology? *Minerva*, 22, 160–82.

Klemm, F. (1959) *A History of Western Technology*. London: Allen and Unwin.

Kline, R. (1987) Science and engineering theory in the invention and development of the induction motor, 1880–1900. *Technology and Culture*, 28, 283–313.

Kuhn, T.S. (1962) *The Structure of Scientific Revolutions*. Chicago: University of Chicago Press.

Latour, B. (1990) Are we talking about skills or about the redistribution of skills? Abstract of paper delivered at a conference on Rediscovering Skill in Science, Technology and Medicine, 14–17 September 1990, University of Bath.

Layton, D. (1973) *Science for the People. The Origins of the School Science Curriculum in England*. London: Allen and Unwin.

Layton, D. (1975) Science or education? *University of Leeds Review*, 18, 81–105.

Layton, D. (1981) The schooling of science in England, 1854–1939. In Macleod, R. and Collins, P. (eds) *The Parliament of Science. The British Association for the Advancement of Science, 1831–1981*. Northwood, Middlesex: Science Reviews Ltd., 188–210.

Layton, D. (1990) Student laboratory practice and the history and philosophy of science. In Hegarty-Hazel, E. (ed.) *The Student Laboratory and the Science Curriculum*. London and New York: Routledge, 37–59.

Layton, E. (1977) Conditions of technological development. In Spiegel-Rösing, I. and de Solla Price, D. (eds) *Science, Technology and Society. A Cross-Disciplinary Perspective*. London: Sage Publications, 197–222.

Layton, E. (1987) Through the looking glass, or news from Lake Mirror Image. *Technology and Culture*, 28, 594–607.

Layton, E. (1990) The nature of technological knowledge. Paper read at the International Conference on Technological Development and Science in the 19th and 20th centuries, University of Technology, Eindhoven, The Netherlands, 7 November 1990.

Levy, E. (1989) Judgment and policy: the two-step in mandated science and technology. In Durbin, P.T. (ed.) *Philosophy of Technology: Practical, Historical and Other Dimensions*. Dordrecht: Kluwer Academic Publishers, 41–59.

Mayr, O. (1976) The science–technology relationship as a historiographic problem. *Technology and Culture*, 17, 663–73.

Miller, W.A. (1857) Report of the President and Council. *Quarterly Journal of the Chemical Society*, 10, 180–91.

Pérez-Ramos, A. (1988) *Francis Bacon's Idea of Science and the Maker's Knowledge Tradition*. Oxford: Clarendon Press.

Rankine, W.J.M. (1857) *The Science of Engineering*. London: Griffin.

Rankine, W.J.M. (1872) *A Manual of Applied Mechanics*, 6th edn. London: Griffin.

Ravetz, J.R. (1985) The methodology of citizen's science. Unpublished paper.

Russell, C. (1983) *Science and Social Change 1700–1900*. London: Macmillan.

Schön, D.A. (1983) *The Reflective Practitioner. How Professionals Think in Action*. London: Temple Smith.

Schön, D.A. (1987) *Educating the Reflective Practitioner*. San Francisco: Jossey-Bass Publishers.

Skolimowski, H. (1966) The structure of thinking in technology. *Technology and Culture*, 7, 371–83.

Staudenmaier, J.M. (1985) *Technology's Storytellers. Reweaving the Human Fabric*. Cambridge, MA and London: Society for the History of Technology and the MIT Press.

Thackray, A. (1976) Discussion of the paper by Robert Multhauf. *Technology and Culture*, 17, 645.

Travis, A.S. (1990) Perkin's mauve: ancestor of the organic chemical industry. *Technology and Culture*, 31, 51–82.

Vincenti, W. (1982) Control-volume analysis: a difference in thinking between engineering and physics. *Technology and Culture*, 23, 145–74.

Weingart, P. (1978) The relation between science and technology – a sociological explanation. In Krohn, W., Layton Jr., E.T. and Weingart, P. (eds) *The Dynamics of Science and Technology*. Dordrecht, Holland: D. Reidel Publishing Company, 251–86.

Whewell, W. (1834) Address to the meeting, in British Association for the Advancement of Science. *Report 1833*. London: John Murray, xi–xxvi.

Wynne, B. (1988) Unruly technology: practical rules, impractical discourses and public understanding. *Social Studies of Science*, 18, 147–67.

1.3

Some Differences Between Science and Technology

J. Sparkes

◼ Introduction

It is quite common for people to talk about 'science and technology' as if it was one thing with a double-barrelled name. For example, people refer to the importance of 'science and technology' for the economic future of Britain, and the government sets up committees to consider what should be done about the decline of 'science and technology'. The reason why people combine these two words in this way is partly because they have several features in common, but mainly because the distinctions have got a bit lost through neglect and from the repeated use of the phrase 'science and technology'. This article is concerned with spelling out once again the distinctions between them.

A rather more insidious practice is speaking of 'technology' as if it were no more than 'applied science'. Indeed some university Engineering or Technology departments teach mainly applied science. Others, more honestly, also do this but call themselves Applied Science departments. Yet others actually teach technology or engineering, but still call themselves Applied Science departments. So there is some confusion around. In this article I want to explain that there are important differences between science and technology, even though they also overlap in an area which might legitimately be called applied science, as indicated in Figure 1.

A further factor leading to confusion is that although there are distinctions between 'science' and 'technology', the distinction between 'scientists' and 'technologists' is much less clear. That is, scientists often make and design things in order to do their experiments, which is a part of technology; and technologists often carry out scientific investigations *en route* to designing a new machine or system. So, while there are definite distinctions between the *subjects* of science and technology, *people* often combine both activities. The fact that Lord Raleigh was a brilliant scientist as well as a brilliant engineer does not mean that there is no distinction between these two

25

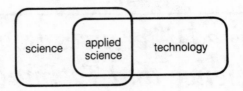

Figure 1 The relationship between science and technology. The two fields overlap but also have distinct characteristics of their own.

areas of activity. Indeed, a word such as 'polymath' only has meaning if distinctions can be made between the different branches of learning in which an individual is an expert.

One final point by way of introduction: some people draw a distinction between 'engineering' and 'technology'. It is sometimes said, for example, that technology is solely concerned with making things, whereas engineering is not only concerned with making things and systems, but is also concerned with wider issues of environmental impact and social consequences too. But then the next person you discuss the matter with will acknowledge that such a distinction can be drawn but will interchange the names – maintaining that technology is the broader activity. I have found that this is an issue which cannot be resolved, so I shall not discuss it further. I shall usually refer to engineering/technology as 'technology'.

■ The goals of science and technology

The *goal of science* is the pursuit of knowledge – often, though not necessarily, for its own sake. Indeed the word 'science' comes from the Greek word for knowledge. Many academic subjects other than science, such as history or literature, are also concerned with the pursuit of knowledge of one kind or other, but science has a particular way of doing it.

- Firstly, it only accepts evidence or data which, at least in principle, could be obtained by anyone at any time, so that subjective judgements of evidence are reduced as far as possible. In the physical sciences they are reduced almost to zero. Indeed, measurements are nowadays often made automatically by machines rather than by people.

- Secondly, science uses logic and mathematics whenever possible to draw conclusions from its theories and data. Again, the room for rival opinions is reduced to a minimum by testing the predictions that each theory leads to in order to try to eliminate those that lead to false conclusions.

The reason for these self-imposed constraints is simply that it is in search of *objective* knowledge. It uses this method to try to bridge the gap between our necessarily subjective view of the universe, and the picture we all have of the existence of a real, external world around us.

For example, the fact that time passes quickly when we are happy and slowly when we are bored is not regarded as evidence about *time* because people have different opinions about it. It could, however, be accepted as evidence about *people* (in the science of psychology) because it is an observation which is experienced by everyone. Or again, as regards the use of logic, science does not accept sweeping generalizations (e.g. 'if you've met one Frenchman you've met 'em all!') or intuitive judgements (e.g. York Minster was struck by lightning because the Almighty did not approve of the appointment of the present Bishop of Durham); it requires that generalizations can always be confirmed and that there is supporting evidence for theories.

The *goal of technology*, by contrast, is to create successful artefacts and systems to meet people's needs or 'wants'. People have a variety of needs, and an even wider variety of 'wants', so designs usually have to meet a number of criteria if they are to be successful. These include, for the customer:

- good quality
- reliability
- good servicing arrangements
- not too expensive
- attractive appearance
- convenient

etc.

The manufacturers or producers are mostly customers of their suppliers, and so have all the above needs too, but they also have needs of their own, including:

- profit
- stable and continuous work
- safe working conditions

Nowadays customers are usually prepared to pay for quality, so price is no longer the key selling parameter that it once was. Successful industries now give quality a higher priority than low cost.

Evidently the goals of science and technology are very different. Nevertheless technology makes much *use* of science, and science makes much use of technology in achieving their respective goals.

▪ Key processes in science and technology

In *science*, the two key processes are *experimentation* and *theory creation*.

Experimentation is mostly for the purposes of discovery of one kind or another:

- it may be discovery of a new fundamental particle such as a 'quark';
- it may be experimentation in a pathology laboratory in a hospital to find out if a patient has diabetes or too much cholesterol;
- it may be concerned with testing steel to ensure it has the right composition to act as a motor-car bearing for 20 years or more.

Many scientists in industry or hospitals or government establishments are engaged in the kind of measurements and testing that the last two examples illustrate. But there are also scientists in industry, universities and research establishments who are mostly seeking *new* and general knowledge about the world or the universe rather than specific knowledge about individual people or products. Scientists, like Newton or Einstein or Maxwell, formulate new theories to enable us to better understand the universe we live in, but they are rare birds, and rightly become recognized by society as a whole.

The creation of theories usually arises out of experimental results, but is more than just making generalizations. Boyle's Law — which states that the volume and pressure of a gas are inversely proportional to each other — is not a theory, it is a generalization; it doesn't explain anything. It merely states concisely the results of many experimental observations. The role of theories is often to *explain* such repeatable phenomena. Boyle's Law, for example, can be explained by the 'Kinetic Theory of Gases'. The fact that apples fall to the ground but the Moon does not is explained by gravity and the concept of inertia.

If it is possible, by a process of *logical deduction* from *relevant theories* and *initial data* to predict correctly new events, science claims to have *understood* these particular events. In addition, each successful prediction strengthens the acceptability of the theory that led to the prediction.

It is better to claim that 'the acceptability of the theory is strengthened', rather than that 'the theory is shown to be true', because a theory can never be completely proved. This is partly because it is always possible that the next time a theory is tested its predictions will turn out to be wrong, and partly because a better theory might be discovered or invented. Newton's Laws of Motion superseded those of Aristotle, and Einstein's superseded those of Newton; but this was not because the later theories were truer, it was because they were more general and could predict certain events more accurately. Einstein's ideas may themselves be superseded sometime. It is true that scientists are *searching* for objective truth, but it can never be demonstrated to have been found. All the same, though it is too grand a claim, it is common enough in science to say that things are 'true' as soon as they are (almost) universally accepted!

If the predictions derived from data and theories do turn out to be false, something must be done. Such a breakdown in the scientific edifice cannot be swept under the carpet. First, the data need to be confirmed – preferably by an independent observer. Then, if the data are confirmed, the theory has to be modified, or additional theories have to be formulated, or the original theory must be abandoned. For example:

- It was once thought that when something burned, a substance called phlogiston was given off. Flames were one manifestation of the phlogiston escaping. Flammable substances were those that contained phlogiston. It was soon found however that the weight of wood ashes, including the smoke etc., was greater than that of the original wood; which meant that phlogiston given off must have a negative weight! This was too much for most scientists to accept, although there is no 'law' against negative weight. Indeed Priestley, the discoverer of oxygen, believed in phlogiston – negative weight included – until he died. (After all, if it isn't phlogiston, what is it that enables hot-air balloons to fly?) Following Lavoisier, burning is now thought of as a process of oxidation.

This is an example of a theory – the phlogiston theory – being abandoned. History is full of such theories though they are rarely referred to in textbooks. Much, however, can be learned about science by studying these early attempts at making sense of the world. An example of a theory being added to is the following.

- Gravitation predicts that all bodies will be found to exert a force of attraction on each other, like the force towards the ground of a weighty object. But when two iron objects are found to repel one another, gravitation is not abandoned. Instead a new theory of magnetism is *added* which in some circumstances gives rise to stronger forces than gravitation.

Theories are rarely logical conclusions derived from the data which stimulated them. In other words, their creation is an imaginative act rather than a reasoned conclusion. New concepts – which lie at the heart of new theories – are imaginative creations designed to explain previously inexplicable phenomena. For example:

- Einstein's Theory of Relativity did not *follow* from the failure of experiments to measure the absolute velocity of the earth through space, although it might have been stimulated by them. His theory arose out of a realization that there was nothing in space to which velocity measurements could relate. Space, or the 'aether' as it was called, in effect had no 'hooks' to which you could attach a measuring device. But, once the implications of this realization were worked out, it enabled the anomalous orbit of Mercury, and several other phenomena, to be explained.

- Newton's Theory of Motion stated that it was 'natural' for bodies to move unaided at uniform speed in a straight line, and so such motion needed no explanation. But neither Newton, nor anyone else, has ever seen a body moving uniformly in a straight line without a force acting on it — they always slow down or move along curved paths unless a force acts on them to overcome friction or the effect of gravity or some other force. So Newton's first 'Law' is not really a 'law' like Boyle's Law, because it is not a generalization from experience; it is an imaginative leap beyond experience which enables experience to be better predicted — and hence explained.

In *technology*, the activities which correspond to theory creation in science are the *design, invention and production of artefacts and systems*.

Thus, in technology, creativity plays a central role rather than an occasional one mostly achieved by geniuses, as in science. Every artefact that is constructed, made or manufactured has first to be designed and possibly invented:

- from pencils to printing presses;
- from carbon steel to computers and cars;
- from tea bags to transistors, trousers and typewriters;

and so on.

Production methods too have to be designed, and are often very ingenious:

- how are tennis, golf and snooker balls made so perfectly round?
- how can biros be made so cheaply?
- how can computers be made to be so reliable?

It is through design and invention and in measuring performance that science enters into technology. The use of scientific theories and data enable the performance of proposed designs to be calculated before the design is implemented. Whenever possible, of course, prototypes are also produced, tested scientifically and evaluated before production starts, but good design is not just a matter of trial-and-error, or even trial-and-success; it is primarily a matter of understanding what one is doing. But science is not the only important input to good design, as is discussed further in a moment.

With 'big' technology, of course, testing before production is not always possible.

- The only way of testing the first 2000 megawatt power station is to build it and to load its output with a city the size of Sheffield!
- One can't do trial landings of a man on the moon without actually putting one there!
- Every bridge is to some extent a special design: only models of it can be tested in simulated hurricanes. Here again science plays its part in

designing test conditions which accurately simulate in miniature the real environment.

There can be many successful design solutions to one requirement, whereas, in science, it is believed that there is only one 'true' explanation of each phenomenon.

■ Routine processes in science and technology

Far more scientists are employed in routine testing than in theory formation. They are employed in pathology labs testing blood samples etc., or in industry monitoring the quality of materials used in production, and so on. Similarly, far more technologists are employed in producing and maintaining existing artefacts and systems than are employed in designing new ones. The difference is that while there are really rather few scientists making significant advances in scientific theories, there are quite large numbers of technologists engaged in design work.

■ Reductionism and holism

Science progresses by the process of *reductionism*, which is the business of distinguishing and defining distinct, separate concepts which, put together, can be said to account for the complex behaviour of the world we live in. For example:

- Matter is broken down into molecules and atoms; which are then broken down into protons and neutrons and electrons; which are nowadays broken down even further. This has been necessary to find explanations of all the phenomena revealed by scientific experiments on matter.
- Various kinds of forces are distinguished so that each can be applied appropriately in calculations of the movements of particular objects or of the stresses in particular structures. Observations are explained as the summed result of all the forces that may be acting, such as gravitation, electric and magnetic forces, electro-magnetic forces, elastic or frictional forces, wind or liquid pressures and so on.

This search for distinct and irreducible concepts, such as separate particles and forces, is the part of scientific method called 'reductionism'. Events are then said to be *caused* by the interplay of these concepts. But it should be noted that not everything can be explained *causally*, so there are limits to the scope of this kind of scientific approach. Complex natural systems that contain a lot of feedback, such as the environment, living creatures, and even the weather are not very amenable to 'causal' explanations, so rather different approaches may be needed. Recently the effects of positive feedback processes have attracted scientific interest, although they have been harnessed in technology for many years.

The process of design in technology involves both *scientific analysis*, as already explained, followed by the reverse process of 'putting it all together again'; a process referred to as '*holism*'.

Thus design in technology involves taking account not only of the implications of scientific predictions, but also of customer requirements, of production and productivity requirements, of the possible effects on the environment, of the implications regarding health, reliability and safety, of the maintenance requirements in case of breakdown, and so on. Holism is the successful integration of all these factors. It is the basis of success in technology.

There is a penalty for having to take so many factors into account in technological designs. Simplified models of the different aspects of a design have to be used, if they are all to be taken into account. The reductionism of science makes it possible to deal with one problem at a time in a laboratory, and so to pursue accuracy as far as experimental techniques will permit. But artefacts and systems usually have to function in uncontrolled environments, so a variety of factors have to be considered together. This can usually only be done if simplified models of each variable factor are used. Good designers are those who choose valid yet simple models, and can thereby entertain many factors simultaneously. Nowadays appropriately programmed computers can help a good deal by making it possible to use more accurate models and still integrate them into a viable whole.

■ Value judgements

Science, in the pursuit of objectivity, excludes, as far as possible, all subjective descriptions of events. As a result it tends to exclude *value judgements* too. This often creates awkward dilemmas for scientists – who are also human beings.

- The discovery of nuclear fission was a triumph of science, but the inevitability of it being used in bombs has caused much heart-searching among scientists.

- The disinterested research into the properties of nerve gases can conflict, in the minds of those who oppose war, with their possible uses.

- Experimentation on animals for the purpose of finding the causes of human disease again sets up conflicts of values.

In technology, however, making value judgements is an inherent part of the design process. No design is perfect; value judgements have to be made all the time as to which solution to a problem is likely to be better, bearing in mind the cost, the appearance, safety, reliability, etc. The choice engineers have to make, or the choice which is sometimes made for them, is usually a choice between two evils; for example:

- Is it better to deny people the cars they want in the belief that they will want the consequences of too many cars even less?
- Is it better to harm the landscape with electricity pylons or to bury the cables and make people pay a lot more for their electricity?
- How much safety in roads, cars, trains, aircraft, etc., is it worth paying for when the risks are small, knowing that nothing can ever be wholly safe?

Value judgements are *always* a part of technology – though they are sometimes ignored by technologists. Value judgements are *not* part of science, even though they concern most scientists.

■ Conclusions and decisions

It is here that differences between science and technology tend to produce differences in attitudes between scientists and technologists. Faced with a problem, scientists, trained in the ways of science, make sure of their facts and use well-tried theories in drawing their conclusions. If there is insufficient data or if there is not yet sufficient understanding of the problem under consideration, the inclination of scientists is to obtain more data or embark on a program of research in order to enable them to come to correct conclusions.

By contrast, technologists, used to having to meet deadlines, learn to make decisions based on incomplete information, bearing in mind that corrective decisions can often be made later if necessary. Indeed, technologists can become impatient with those who put off a decision until more data has been obtained, since there are rarely perfect solutions to the conflicting requirements of most technological problems. The hope is, in delaying a decision, that additional information or better theories will reduce the decision to a clear, logical conclusion; but the contrary is more often the case. The additional information only emphasizes the differences between the possible alternative decisions. It is often better, except perhaps in life-threatening situations, to make a workable decision, and to make corrections later if necessary, than to delay matters in the hope of making a better decision later.

So this distinction can be summarized as being the distinction between 'drawing conclusions' and 'reaching decisions'. Not surprisingly, academics, many of whom are engaged in the search for knowledge, are often bad at making decisions. Equally, industrialists sometimes rush to a decision when a little more information would enable them to come to a valid conclusion or at least make a better decision.

It is noticeable that 'scientists' and 'technologists' can be expected to react differently when things go wrong. If the scientific predictions of future events turn out to be mistaken, scientists check calculations, check data, and reappraise the scientific theories used, since there must be a cause for the mistake! Technologists, on the other hand, already know that their decisions may not be wholly justified logically, and that therefore there is some risk in taking them. So, if a decision turns out badly, technologists' first inclinations are to take further decisions aimed at rectifying the situation. But they would also look for reasons for the failure of the first decision, and perhaps formulate some research proposals on the matter for their scientific colleagues to work at.

■ Research

Finally, because of their different goals, science and technology engage in different kinds of research. A good deal of technology is concerned with development – that is, working up an idea for a new artefact or system so that it will be successful on the open market. But this is not research.

In science, 'research' usually means the search for new knowledge and understanding. But it can also mean, and usually does in other fields, the gathering together, and re-interpretation, of existing but dispersed knowledge relevant to a particular problem. This is not the kind of research I want to discuss here, where I am concerned about the differences in the nature of new knowledge looked for in science and technology.

In the case of science, the search is usually for new data – often obtained through controlled experiments – and for *causal explanations* of them. The explanations may take the form of:

- a theory of gravitation to explain why apples fall, but the moon does not;
- a double helix structure for DNA to explain how characteristics are passed on from cell to cell;
- the concept of the Big Bang to explain how the universe began.

In technology the search is for the principles underlying *better processes*, or for better ways of making or doing things. The findings may take the form of:

- the principles behind negative feedback as a means of obtaining stable, self-correcting processes;
- different techniques for storing information in computers so that it can readily be accessed;
- a machine which can recognize human speech, and can type out what you say to it as you say it;
- jet engines as an improvement on piston engines for aircraft.

Of course, each of these technological research findings may well make use of scientific theories and results, just as a great deal of scientific research has reason to use the latest technology to obtain some of its experimental results. This does not mean that science and technology are the same, or even complementary. It simply means that, although each has its special characteristics, they make use of each other.

■ Topics covered in science and technology courses

A different kind of difference between science and technology concerns the range of topics covered in courses on these subjects. Since technology can have such an effect on the environment, on people's lives and on social issues, it is quite appropriate to include consideration of such topics in technology courses. But since many technological developments depend on advances in science it is *also* quite appropriate to teach aspects of science in technology courses. However, it is very difficult to deal with these two 'ends' of the subject in the same course. So what should one do in planning a course? One solution is as follows.

It is possible to construct a kind of hierarchy or 'ladder' of topics within one subject area, with the top at the general, social level, and the bottom at the most analytic scientific level. In a subject like electronics, the hierarchy might look like this:

(1) Impact on people, society and the environment of electronic artefacts, production methods, etc.

(2) Descriptions of complete electronic artefacts, such as TV, radar, hi-fi, computers, etc., and an analysis of their properties.

(3) System analysis and system design of complete artefacts such as those in (2), including costs, quality, reliability, etc.

(4) Analysis and design of the constituent elements of complete artefacts (e.g. amplifiers, integrated circuits of various kinds, memories in computers, etc.).

(5) The properties of circuit components used in the constituent elements (e.g. transistors, capacitors, inductors, diodes, transformers, etc.).

(6) Scientific explanations of the properties of the components listed in (5).

(7) The scientific principles and theories upon which the explanations in (6) depend.

The problem in the design of technology courses is deciding how much of this ladder or hierarchy to try to cover in a single course. It turns out that it is very difficult to construct a good course which deals satisfactorily (i.e. not very superficially) on more than about three adjacent rungs on such a ladder. That is, it is a good idea to focus on one rung of the ladder, and deal with the important relevant concepts that apply to it; but also to give these facts and concepts the *context* provided by the next rung up, plus the degree of *analysis* provided by the next rung down.

■ Conclusion

All the above differences between science and technology are summarized in Table 1. However, it should not be forgotten that science and technology also have similarities. For example, they both make use of mathematics and modelling. This makes it possible, for instance, for scientists and technologists to learn each other's jobs without too much difficulty, if they have a mind to do so.

Table 1   Some differences between science and technology.

SCIENCE (Goal: the pursuit of knowledge and understanding for its own sake)	TECHNOLOGY (Goal: the creation of successful artefacts and systems to meet people's wants and needs)
Key scientific processes	*Corresponding technology processes*
Discovery (mainly by controlled experimentation)	Design, invention, production
Analysis, generalization and the creation of theories	Analysis and synthesis of designs
Reductionism, involving the isolation and definition of distinct concepts	Holism, involving the integration of many competing demands, theories, data and ideas
Making virtually value-free statements	Activities always value-laden
The search for, and theorizing about, causes (e.g. gravity, electromagnetism)	The search for, and theorizing about, new processes (e.g. control; information; circuit theories)
Pursuit of accuracy in modelling	Pursuit of sufficient accuracy in modelling to achieve success
Drawing correct conclusions based on good theories and accurate data	Taking good decisions based on incomplete data and approximate models
Experimental and logical skills	Design, construction, testing, planning, quality assurance, problem-solving, decision-making, interpersonal and communication skills
Using predictions that turn out to be incorrect to falsify or improve the theories or data on which they were based	Trying to ensure, by subsequent action, that even poor decisions turn out to be successful

1.4

The Three Rs

B. Archer

[In this article, Bruce Archer argues from an educational point of view, not that technology should be distinguished from science, but that 'design' is distinct from both science and humanities.]

The world of education is full of anomalies. Take that extraordinarily durable expression 'The Three Rs', for example. It is very widely held that when all the layers of refinement and complexity are stripped away, the heart of education is the transmission of the essential skills of reading, writing and 'rithmetic. This expression is internally inconsistent, to begin with. Reading and writing are the passive and active sides, respectively, of the language skill, whilst arithmetic is the subject matter of that other skill which, at the lower end of the school, we tend to call 'number'. So the expression 'The Three Rs' only refers to two ideas: language and number. Moreover, the word 'arithmetic' is mispronounced as well as misspelled, giving the impression that the speaker takes the view that the ability and the necessity to do sums is somehow culturally inferior. If challenged, most who use the expression would deny they intended any such bias, but aphorisms often betray a cultural set.

Explicit or implied denigration of science and numeracy in favour of the humanities and literacy was certainly widespread in English education up to and beyond the period of the second world war, and was the subject of C.P. Snow's famous campaign against the separation of 'the two cultures' in 1959. The two cultures may be less isolated from one another these days, and may speak less slightingly of one another, but the idea that education is divided into two parts, science and the humanities, prevails. There are many people, however, who have always felt that this division leaves out too much. Art and craft, dance and drama, music, physical education and sport are all valid school activities but belong to neither camp. There is a substantial body of opinion, not only among teachers but also among groups outside that profession, which holds that modern society is faced with problems – such as the ecological problem, the environmental problem, the quality-of-urban-life problem, and so on – all of which demand of the population of an affluent industrial democracy, competence in something else besides literacy and numeracy. Let us call this

37

competence 'a level of awareness of the issues in the material culture' for the time being. Under present circumstances, it is rather rare for a child who is academically bright to take art or craft or home economics or any of the other so-called 'practical' subjects having a bearing on the material culture to a high level in the fourth, fifth or sixth forms. Universities and professional bodies do not usually accept advanced level qualifications in these subjects as admission qualifications for their courses, even where the course, such as architecture, engineering, or even, in some cases, art and design, is itself concerned with the material culture. It is really rather an alarming thought that most of those who make the most far-reaching decisions on matters affecting the material culture, such as businessmen, senior civil servants, local government officers, members of councils and public committees, not to mention members of parliament, had an education in which contact with the most relevant disciplines ceased at the age of thirteen.

■ A third area in education

The idea that there is a third area in education concerned with the making and doing aspects of human activity is not new, of course. It has a distinguished tradition going back through William Morris all the way to Plato. When St Thomas Aquinas defined the objects of education in the 13th century, he adopted the four cardinal virtues of Plato (prudence, justice, fortitude and temperance) and added the three Christian virtues (faith, hope and charity). These have a quaint ring in modern English, but Plato's virtues, rendered into Latin by St Thomas Aquinas, were taken to mean something quite specific and rather different from their modern English interpretations. To St Thomas Aquinas, *prudentia* meant 'being realistic, knowing what is practicable'; *justitia* meant 'being ethical, knowing what is good'; *fortitudo* meant 'being thorough, knowing what is comprehensive'; *temperentia* meant 'being economic, knowing when to leave well enough alone'. It is no coincidence that in our own day Dr E.F. Schumacher, in the epilogue to his book *Small is Beautiful*, quotes the four cardinal virtues of Plato as the basis for the socially and culturally responsible use of technology in the modern world. Certainly the craft guilds, who bore a major responsibility for the general education of the populace following the Renaissance, took the view that a virtuous education meant learning to know what is practicable, what is good, what is comprehensive, and what is enough, in a very broad sense. It is a curious twist in fortunes that when the craft guilds lost their general educational role somewhere between the 14th and 18th centuries, it was the rather narrow, specialist, bookish universities, academies and schools which had been set up to train priests to read and translate the scriptures which became the guardians of what we now call general education. No wonder our education system came to be dominated by the humanities.

When Sir William Curtis, MP, coined the phrase 'The Three Rs' in or about 1807, he placed an emphasis on literacy which reflected the virtual monopoly that the church then had in the running of schools. I had an old great-aunt who protested fiercely whenever the phrase 'The Three Rs' was mentioned. She swore that Sir William had got it all wrong. The Three Rs were:

(1) Reading and writing.
(2) Reckoning and figuring.
(3) Wroughting and wrighting.

By wroughting she meant knowing how things are brought about, which we might now call technology. By wrighting she meant knowing how to do it, which we would now call craftsmanship. From reading and writing comes the idea of literacy, by which we generally mean more than just the ability to read and write. Being literate means having the ability to understand, appreciate and value those ideas which are expressed through the medium of words. From reckoning and figuring comes the idea of numeracy. Being numerate means being able to understand, appreciate and value those ideas that are expressed in the language of mathematics. It was from literacy that the rich fabric of the humanities was woven. It was from numeracy that the immense structure of science was built. But what of wroughting and wrighting? It is significant that modern English has no word, equivalent to literacy and numeracy, meaning the ability to understand, appreciate and value those ideas which are expressed through the medium of making and doing. We have no word, equivalent to science and the humanities, meaning the collected experience of the material culture. Yet the output of the practical arts fills our museums, and galleries, equips our homes, constructs our cities, constitutes our habitat.

Anthropology and archaeology, in seeking to know and understand other cultures, set at least as much store by the art, buildings and artefacts of those cultures as they do by their literature and science. On the face of it, if the expression of ideas through the medium of doing and making represents a distinctive facet of a culture, then the transmission of the collected experience of the doing and making facet should represent a distinctive area in education.

■ The vacant plot

If there *is* a third area in education, what distinguishes it from science and the humanities? What do science and the humanities leave out? It now seems generally agreed among philosophers of science that the distinctive feature of science is not the subject matter to which the scientist turns his attention, but the kind of intellectual procedure that he brings to bear upon it. Science is concerned with the attainment of understanding based upon observation, measurement, the formulation of theory and the testing of theory by further observation or experiment. A scientist may study any phenomenon he chooses; but the kind of understanding he may achieve will be limited by the observations he can make, the measures he can apply, the theory available to him and the testability of his findings. Some sorts of phenomena may therefore be inappropriate for scientific study, for the time being or for ever. Some sorts of knowledge will be inaccessible to science, for the time being or for ever. Moreover, the scientist is concerned with theory, that is, with generalizable knowledge. He is not necessarily competent or interested in the practical application of that knowledge, where social, economic, aesthetic and other considerations for which he does not possess any theory may need to be taken into account. He would regard most of the making and doing activities of the material culture as outside his scope, although he would be prepared to bring a scientific philosophy to bear upon the study of the making and doing activities of other people.

Among scholars in the humanities there seems to be less agreement about the nature of their discipline, apart from unanimity in the view that it is quite distinct from science. There is a fair consensus that the humanities are especially concerned with human values and the expression of the spirit of man. This justifies scholars in the humanities in studying the history and philosophy of science, but not in contributing to its content. There also seems to be a measure of agreement, by no means universal, that the humanities exclude the making and doing aspects of the fine, performing and useful arts, although their historical, critical and philosophical aspects would still be fair game for the humanities scholar. It is interesting to note that writers on the science side frequently mention technology and the useful arts as being excluded from their purview, presumably because they are only just outside the boundary. Writers on the humanities side frequently mention the fine and performing arts as being excluded, presumably because they, too, are only just outside. A third area in education could therefore legitimately claim technology and the fine, performing and useful arts, although not their scientific knowledge base (if any) or their history, philosophy and criticism (if any) without treading on anyone else's grass.

■ The naming of the parts

Clearly, the ground thus left vacant by the specific claims of science and the humanities extends beyond the bounds of 'the material culture' with whose pressing problems we began. The performing arts are a case in point. There are other areas, such as physical education, which have not been mentioned at all. It would be tempting to claim for the third area in education everything that the other two have left out. However, we should stick to our last, if I may take my metaphor from the doing and making area, and clarify the question of education in the issues of the material culture.

Any subject which relates with man's material culture must necessarily be anthropocentric. A discipline which claims, as some kinds of science do, to deal with matters that would remain true whether man existed or not, would be ruled out from our third area. Material culture comprises the ideas which govern the nature of every sort of artefact produced, used and valued by man. Those ideas which take the form of scientific knowledge would belong to science. The historical, philosophical and critical ideas would belong to the humanities. What is left is the artefacts themselves and the experience, sensibility and skill that goes into their production and use. If the human values, hopes and fears on which the expression of the spirit of man are based are shared with the humanities, the striving towards them, and the inventiveness that goes into the production and use of artefacts, is a necessary characteristic of our third area. Any discipline falling into this area must therefore be aspirational in character, and, to take them clearly out of both the science and the humanities fields, it must be operational, that is to say, concerned with doing or making.

Under these tests, how do the subjects ordinarily left out by the traditional science/humanities division fare? The fine arts, which in schools can be executed in a variety of materials such as ceramics and textiles as well as through the medium of painting and sculpture, clearly fall into the third area. In the useful arts, woodwork and metalwork would usually qualify. Technical studies are sometimes conducted in such a way that they are not actually concerned with doing and making, and therefore may or may not rank as science, instead. Similarly, environmental studies might or might not fall into the third area, according to their manner of treatment.

Home economics presents a problem. Taken as a whole, home economics is clearly anthropocentric, aspirational and operational, and therefore falls centrally into the third area. In practice, however, home economics may be taught in schools through the medium of individual subjects ranging from needlecraft taken as fine art through home-making taken as useful arts to nutrition taken as science. So home economics, too, may fall into science, the humanities or the third area, according to the manner of treatment adopted.

Outside the bounds of the material culture altogether are the other subjects explicitly left out by the first and second areas. Among the performing arts, music might qualify as anthropocentric, aspirational and operational. So might drama and perhaps dance. So might gymnastics, the way it is pursued these days, but probably not the other areas of physical education. But this is going too fast. Any number of objections can be raised and counter-arguments offered in respect of many, but perhaps not all, the subjects I have mentioned as belonging or possibly belonging to an alleged third area in education. The point I wanted to make is simply this. The justification for the nomination of a third area in education lies not in the existence of subjects which do not fit readily into the definitions of science and the humanities, but in the existence of an approach to knowledge, and of a manner of knowing, which is distinct from those of science and the humanities. Where science is the collected body of theoretical knowledge based upon observation, measurement, hypothesis and test, and the humanities is the collected body of interpretive knowledge based upon contemplation, criticism, evaluation and discourse, the third area is the collected body of practical knowledge based upon sensibility, invention, validation and implementation.

■ The naming of the whole

This leaves us with the problem of finding the correct title for the third area. The term 'the arts' would be ideal, if the expression had not been appropriated by, and used more or less as a synonym for, the humanities. Plato would not have objected to 'aesthetics', but that has taken on a special and distracting meaning in modern English. 'Technics' has been used, and is in the dictionary, but has not proved very popular in educational or common use. A term which has gained a good deal of currency especially in secondary schools in England and Wales is 'Design', spelt with a big D and used in a sense which goes far beyond the day-to-day meaning which architects, engineers and other professional designers would assign to it. Thus design, in its most general educational sense, where it is equated with science and the humanities, is defined as the area of human experience, skill and understanding that reflects man's concern with the appreciation and adaptation of his surroundings in the light of his material and spiritual needs. In particular, though not exclusively, it relates with configuration, composition, meaning, value and purpose in man-made phenomena.

We can then go on to adopt, as an equivalent to literacy and numeracy, the term 'design awareness', which thus means 'the ability to understand and handle those ideas which are expressed through the medium of doing and making'. The question of the language in which such ideas may be expressed is an interesting one. The essential language of science is notation, especially mathematical notation. The essential language of the humanities is natural language, especially written language. The essential language of design is modelling. A model is a representation of something. An artist's painting is a representation of an idea he is trying to explore. A gesture in mime is a representation of some idea. Everybody engaged in the handling of ideas in the fine arts, performing arts, useful arts or technology employs models or representations to capture, analyse, explore and transmit those ideas. Just as the vocabulary and syntax of natural language or of scientific notation can be conveyed through spoken sounds, words on paper, semaphore signals, Morse code or electronic digits, to suit convenience, so the vocabulary and syntax of the modelling of ideas in the design area can be conveyed through a variety of media such as drawings, diagrams, physical representations, gestures, algorithms – not to mention natural language and scientific notation. With all these definitions in mind, it is now possible to show the relationships between the three areas of human knowledge according to the diagram in Figure 1.

The repository of knowledge in science is not only the literature of science but also the analytical skills and the intellectual integrity of which the scientist is the guardian. The repository of knowledge in the humanities is not simply the literature of the humanities but also the discursive skills and the spiritual values of which the scholar is the guardian. In design, the repository of knowledge is not only the material culture and the contents of the museums but also the executive skills of the doer and maker.

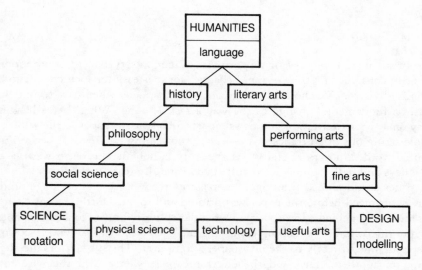

Figure 1   The relationship between the three areas of human knowledge (humanities, science and design).

1.5

Who's Paying the Driver?

I. Lowe

The broad issue of equity in modern Australian society is a most important factor to consider in a time of rapid change, for it is easy for such change to be socially divisive. We should think about what happens when new technology causes a fundamental change to the way a task is done. When I was hacking away on a typewriter producing news reports for a student paper in the 1960s, a whole string of skilled trades was necessary to turn my fractured prose into the printed word. Linotype operators set the story in hot-metal type, it was placed in galleys and galley proofs were pulled and carefully read, sub-editors laid out the pages and designed headings, compositors made up the page, page proofs were scrutinized before the paper was printed and so on. Until a recent change in senior management, I was writing a regular column for a national newspaper. I wrote my column on a laptop computer and used a modem to send it down a telephone line, directly to the newspaper's computer. The sub-editors did their thing on another screen and the electronic mess went down the wire to a typesetting machine. The words didn't appear on paper until the newspaper was printed.

Now think about how the changes have affected different people. For the writer or the sub-editor, there has been little fundamental change. The job may have become a little more satisfying; it is certainly easier to revise a story and get it right if you can play around with it on the screen, rather than screwing up the first piece of copy paper and trying again. Now think what has happened to the jobs of the linotype operator or the compositor, those skilled artisans who used to produce the metal form of the words from which the paper was printed.

Those jobs no longer exist. In terms of the workforce of a newspaper, those people are not employed any more. There are other examples; those who used to work in service stations pumping petrol usually lost their jobs when the self-service pumps were installed, for example. If you think that this problem only affects the unskilled or blue-collar workers, reflect on the revolution which has hit banks in recent years. Like I do, you probably often forget to go to the bank on Friday, but the day is saved by the automatic teller which will give you folding money at 11 p.m., or early Saturday on your way to do the shopping.

You, as a consumer, have benefited from the introduction of the automatic teller machines. Look around your bank branch some time, and compare the number of people behind the counters with the workforce of five or ten years ago. The banks no longer employ large numbers of bright young school leavers to begin work on the counter dealing with customers and gradually work their way up through the system. Computer experts are employed, but the banks are no longer recruiting large numbers of school leavers, so a steady stream of traditional white-collar jobs has dried up.

[Of] [. . .] the variety of forces which motivate the introduction of new technology [social equity is] not one of them. Indeed, the most common motivations for new technology, such as technical advancement or economic gain, make no pretence to incorporate the equity dimension. Whether or not a new technology is accepted is usually determined by the market, by the decisions of individual consumers and corporate groups.

Economists tend to believe in markets because they think that they are the most efficient means of allocating resources between competing interests. Whether that is true or not, even the most besotted admirer of markets would probably not claim that they contain any feature or mechanism which could promote equity, even in principle. Indeed, there is a fundamental sense in which markets will always promote inequity. If a market is used to allocate a scarce resource, whether it is beach-front land or accomplished professional footballers, those with large bank accounts will always be able to outbid those with lighter wallets. Thus allocation through a market means that those who are rich will get what they want and those who are poor will miss out. Those of us who are not rich are limited to competing for those things which the rich do not want sufficiently to bid the price up out of our reach.

This is a compelling argument for government intervention in the marketplace. At the economic feeding trough, the strong and aggressive piglets will always push aside those which are scrawny or meek if government abdicates its responsibility to produce some form of order. Leaving things to the market constitutes acquiescence by government in concentrations of power which we know will be abused. It has been wisely said that in any struggle between the powerful and the powerless, to do nothing is not to be neutral but to side with the powerful.

Leaving our economy to the forces of the market is simply an abdication of the historic responsibility of government to restrain the power of those who could otherwise use their economic clout against the interest of the community as a whole. As the divisions between the powerful and the powerless are widened by technological change, there is a greater responsibility on our political and legal system to ensure that power is moderated in the public interest.

It is helpful in reaching an understanding of the economic dimension of technology to go back 200 years to the founder of serious economics, Adam Smith. He has been much maligned because of the simplistic views of some of his modern disciples, but Adam Smith actually had quite a sophisticated understanding of the way the world works. As Tom Fitzgerald pointed out in the 1990 Boyer Lectures, we would probably be much better governed if those countless economic advisers who assist government had actually read all of Adam Smith's key work, *The Wealth of Nations*. This first analysis of technological change was based on the view that there were three key factors needed to produce anything useful enough for people to want to buy it. The three factors were land, labour and capital. In that fundamental sense, his analysis was more complete than the picture used by many modern neoclassical economists, who ignore the vital input from land or natural resources to reach the absurd conclusion that perpetual growth is possible in our closed system.

In the case of technological change, Smith saw the crucial trade-off as being between capital and labour. He thought that entrepreneurs who behaved rationally would work out the balance between two competing ways of carrying out a job: paying wages for human labour or investing capital in a machine. A factory owner will buy a machine if the cost of production is reduced: in other words, if paying interest on the money borrowed to buy the machine costs less than wages for workers to do the same job.

Of course, real life is more complicated than that simple trade-off. The first factories were built to control the workforce rather than to cut costs. Once the workers were in the factories, it was often possible to force them to accept lower wages by the threat of mechanization, thus complicating the trade-off between equipment and labour. In modern Australia, the comparison is weighted toward mechanization by the reprehensible payroll taxes imposed by state governments, since these form an economic penalty for employing people rather than machines.

A further complication is the issue of what happens to the extra productivity when machines are used. Smith saw the increase in output as being universally beneficial. New technology would only be introduced if it made the business more profitable. The extra profits can only be distributed as either higher wages to the workers or larger dividends to those owning the company. In either case, Smith argued, the extra cash would be spent in the community, thus providing extra work for the butcher, the baker and the candlestick maker, so the whole community would benefit. A modern version of this argument was put forward in the Myers Report on technological change in Australia, concluding that the community as a whole would benefit from new technology. The fallacy of applying this argument to our modern economy was pointed out by Professor Ron Johnston, who noted that a significant number of our productive enterprises are owned by overseas interests. That means, in turn, that a significant fraction of the extra wealth produced by technological change does not flow into the community, but instead flows off to the United States or Britain or Japan or wherever the stakeholders happen to be based, benefiting their local economies rather than ours.

The question of the economic role of technology in the modern industrial state has been analysed by the American economist, J.K. Galbraith. He points out that many of the characteristics of modern organizations stem from what he calls the imperatives of technology. He argues that the harnessing of technology to the production of goods and services has irreversibly changed the nature of the firms engaged in those tasks, leading inexorably to an increasing time span for completion of production tasks, an increase in the capital investment required, more specialized labour, an increasingly important role for organization as a feature of production, a greater need for long-term planning and an increasing tendency to want the state to bear major risks.

Galbraith argues that these trends have ended forever the era of large corporations run by one entrepreneur, such as Henry Ford. In the modern era, corporations are controlled by technical experts, who hold between them 'the diverse technical knowledge, experience or other talent which modern industrial technology and planning require'.

The massive investment of capital, equipment and labour in the productive process leads to fundamental changes to the theoretical interactions between producers and consumers in a market. Producers can no longer afford to allow the demand for their goods and services to be determined by the whim of the consumer, and so set out to create and manage demand. It could be claimed that the main purpose of the advertising industry is to ensure that consumers do not make economically rational decisions. Many large corporations, especially in countries such as the USA, rely mainly on military contracts: the ultimate form of insulation from the pressures of market forces and consumer demand. To quantify this effect, US military contracts in 1989 added up to US$129 billion, equal to more than half of the total economic output of Australia.

Large modern corporations typically don't attempt to maximize their profits, but aim instead at corporate survival and growth. The scale of investment and the time scale of planning lead corporations to be risk-averse and seek the support of the state for their activities. As Galbraith put it, 'The decisive power in modern industrial society is exercised not by capital but by organization, not by the capitalist but by the industrial bureaucrat'. He argued that complex organization, and therefore bureaucracy, is inescapable in advanced industrial society.

This analysis was a brave attempt by a leading economist to explain the real world of today; by comparison, simple market economics is of little assistance in understanding the modern technological state. Galbraith's insights are vitally important. We can only see what future options we have if we understand the limitations imposed by the general properties of technology.

First, modern technology requires a highly developed system of division of labour; modern work is increasingly specialized. The general mechanic gave way to specialists like the motor mechanic, who today is likely to concentrate on one area, such as brakes, or one brand of car. Complex tasks are subdivided into jobs suitable for the various specialists.

Second, we have a highly centralized production and distribution system; when I go to the shop I am unlikely to be offered local produce or even Australian produce if the shop has been able to buy cheaper goods from further away. The extreme example of this effect is the spreading network of so-called fast-food outlets, taking pride in producing the same bland taste in Melbourne as in Miami, Milan and even Moscow.

A third characteristic is that the economic system is oriented towards wants, rather than needs; think about the fraction of your income spent on basic needs compared with indulgences such as fashion, entertainment, travel, sport and hobbies.

We now have a regulated market system for goods and services; not only does government control such basics as hygiene and food quality, it also regulates trade practices, consumer rights and even shopping hours. Overall, the state has a greatly increased role. Even in Australia which has, despite the constant claims that we are overgoverned, a smaller government sector than most industrialized countries, it is still true that the state is a major supplier of goods and services and the major purchaser of high-technology products.

One of the most interesting developments in recent years has been the massive growth in world trade, especially in value-added goods and services. Simple commodities account for a relatively small fraction of economic activity. There is a very good reason for this development. Modern technology is more sophisticated and so uses smaller quantities of raw materials to do the same job. Today's computers are much smaller than those of a decade ago, and use smaller amounts of minerals. The same is true of TV sets, cars and a range of other devices in common use. Minerals and other raw materials are of decreasing importance in world trade. Australia is very unusual in having a trade pattern dominated by minerals and agricultural produce, when world trade is dominated by information and highly transformed manufactured goods.

We should also recognize that politics in the complex modern industrial state is mainly concerned with economic management; think about the extent to which the political agenda in Australia is dominated by interest rates, trade figures, wage settlements and the public sector borrowing requirement. Keeping a complex modern economy operating is almost a circus act, so many different factors need to be kept in balance.

Finally, there is widespread use of elaborate structures for management of the process of production and distribution; even such areas as marketing now boast sufficient complexity to merit academic appointments in many of our universities, and graduate schools of management are devoted to the arcane skills of keeping the modern company functioning.

The general point is that modern industrial society is complex and relies on a high level of interconnection, regulation and government intervention. Additionally, most big corporations take great trouble to insulate themselves from the sort of random fluctuations which occur in a market economy. Those who argue for a simple approach of leaving the modern state to market forces are being at least as unrealistic as those who seek a return to a pre-industrial state of Arcadian bliss. Just as we could not hope to feed the current population from an agrarian economy, we could not support our industrial system on the basis of an 18th-century free market.

Galbraith's analysis is an updated version of a theory put forward 50 years ago by James Burnham in his book *The Managerial Revolution*. Burnham saw the world of his time undergoing what he regarded as a social revolution to a new 'managerial society'. He saw the control of corporations being taken away from the owners by the managers, a process which has continued to this day. The nominal owners of large companies – the shareholders – have very little control over what is done. By controlling the flow of information to the nominal owners, the managers retain control of the key investment decisions affecting the future of the corporation.

Burnham foresaw the erosion of national sovereignty which has recently occurred. He argued that a traditional sovereign nation made its own laws, set its own tariffs and other controls on imports and exports, regulated its own currency and set its own foreign policy. He saw all these activities as incompatible with the needs of modern technology such as 'the complex division of labour, the flow of trade and raw materials'. As I have mentioned, world trade and commerce are now dominated by intangibles rather than physical objects. As a consequence, increasing integration of the world economy is a fact of life. National governments can and do regulate the flow of physical commodities into and out of their countries; they erect customs barriers at their borders and tax flows of commodities. But governments have difficulty even detecting the flows of information, let alone regulating these areas of economic activity. It has been said that the political union of Europe in 1992 is merely giving political recognition to what has become a financial reality, which is that the borders between the separate nations of Western Europe have very little significance in today's world.

I want to make one more general point before putting these ideas into a local context and exploring their implications for Australia. The recent dramatic rise in the importance of information technology has changed the world in some fundamental ways.

Where energy technology has been the key to most of the past industrial development, information technology will play a similar role in the future modern society because it has a great capacity to influence other technological developments. Modern information technology has facilitated the implementation of change in a wide range of areas, as well as accelerating the pace of transfer of ideas and developments from one place to another. Not only does this result in more rapid diffusion of new ideas, but it also removes the need for innovations to occur in particular geographical areas. The smokestack industries of the original Industrial Revolution developed where the appropriate resources were located; for example, steel-making was carried out where coal, iron ore and water for transport were close to each other. Newcastle and Wollongong are Australian examples. The industries of today and tomorrow are much less likely to be tethered to physical resources, and much more likely to be located near markets or sources of skilled labour.

Information technology has also fundamentally changed the power relations in society. Large corporations are now equipped with global communications networks, but so are governments and various non-govern-ment organizations, such as conservation groups, indigenous people and political oppositions. The very success of contemporary technology has placed powerful communications capacity within the reach of a mass market. This has led to some unexpected consequences. Let me give you four examples.

Where once only large corporations could afford computers and high-quality printing, those technologies are now within the reach of small companies, community groups or individuals. Second, conservation groups are playing an increasingly important role in the negotiation of global treaties, because their sophisticated use of information systems often means that they are better informed than the diplomats negotiating on behalf of governments. As a third example, indigenous peoples in the Amazon have begun using video to record negotiations with city entrepreneurs to ensure that agreements are honoured. Finally, opposition to the regime of the Shah of Iran was built up despite total control of the mass media, by smuggling tape cassettes of speeches by the Ayatollah Khomeini from mosque to mosque. Thus information technology has become a powerful levelling tool. Where George Orwell foresaw our leaders using video to spy on us, through the media we are now using video to spy on our leaders!

This whole analysis of the characteristics of technological society has implications for Australia at three levels: our internal structure of states and territories, our overall national economy and our relationships with other countries. There are important ways in which our Federal system is a handicap in the modern world. I mentioned that the barriers between the separate nation-states of Europe are becoming less important. By contrast, we still behave largely as if we were a series of warring tribes. Our states have separate legal systems, different tax structures, individual approaches to education, even different road rules. The changes necessary in moving between our various states are often greater than those in moving between the separate nations of Europe. It was recently observed that we are the most competitive nation on earth, as we spend most of our time and energy competing with each other.

We certainly do not present a united or even coordinated approach to the rest of the world. As far as the structures of our national economy goes, we are also out of step with the rest of the industrialized world. In overall terms, we are already what has been called a post-industrial society. About three-quarters of our wealth is produced by the broad area of services, with only about a quarter coming from physical commodities: agricultural produce, minerals and manufactured goods. However, our trade pattern is still heavily oriented toward the export of minerals and farm produce, with a massive trade imbalance in manufacturers. As the value of commodities has been falling in relative terms for at least 40 years, and arguably for most of this century, our terms of trade have inevitably declined.

Conventional economic wisdom has encouraged us down this path by its belief that we should pursue activities in which we have a comparative advantage over other nations. Some people even believe that the solution to our current economic problem consists of exporting more minerals, even though this approach would entrench more firmly a trading pattern characteristic of a Third World country. With our national wagon hitched firmly to the fading star of commodity exports, we would experience continuing economic decline to the only status even worse than the banana republic threatened by Paul Keating: becoming a banana monarchy.

I have mentioned that the world economy is increasingly an international organism. This has a clear effect on economic management. Our government now frames its budget and waits with almost pathetic eagerness to see what the New York money market thinks about it. We think it is a good budget if the verdict is that the Americans want to buy still more of our productive assets. We have little room for manoeuvre as a result of our trade structure and the level of foreign investment that has been encouraged in the past by our short-sighted leaders.

We need to outgrow the short-term economic approach which allows the future to be determined by simple considerations of profit, whatever the long-term consequences for our living standards or even our political independence. We should be making a concerted effort to plan rationally for the future. I can't see that the current pattern of economic activity in Australia makes sense even in economic terms. It has very serious potential social and political consequences. Additionally, it does not appear to be ecologically sustainable.

[. . .] options for avoiding further economic decline [. . .] hinge crucially on the current enthusiasm for becoming what has been called a clever country. Like former Science Minister Barry Jones, I would much prefer that we become an intelligent society, but that is another story.

I want to turn, finally, to the issue of our economic relationships with other countries. Technology being now a powerful economic force, the nations which are doing well economically are those which have successfully harnessed technology to their needs: countries such as Sweden, West Germany and Japan.

Take the case of Japan, a dramatic example of the saying that past trends are not future destiny. Forty years ago, Japan was a middle-rank economic power with a reputation for shoddy goods. Today it is the dominant economic power of the world, with all ten of the world's largest banks and over a third of all the deposits in all the banks of the world. This is precisely because of the way the Japanese economy has specialized in the sort of value-added products which are the key to economic success in the modern world.

During the same period the United States has failed to maintain its technological competitiveness. In fact, the United States has declined in two decades from the richest nation on earth to the largest debtor in human history. The indebtedness of the USA is two to three times the amount per head of the next worst case, Brazil. Its debt position is much worse than ours. Furthermore, the USA is increasingly moving toward the trading pattern of a Third World country, exporting raw materials and buying its elaborately transformed manufactured goods from Japan and other countries of east Asia. The USA has been described, like Australia, as a very rich Third World country.

The changing economic relationship between Japan and the United States should command our attention, as should the growth in the economic influence of other countries of east Asia. As a nation, we are still alarmingly ignorant of the history, the politics, the culture and above all else the diversity of the nations to our north. It is difficult to see a long-term benefit in this strategy.

In the absence of any alternative plan we are likely to see Australia moving more comprehensively into the role of an economic colony of Japan, doing the things which colonies traditionally do: supplying raw materials, acting as a market for manufacturers and providing a suitable site for holidays. If we want to shape our own future rather than have it shaped by others, we need to adopt a different approach; the future is bleak if we continue to pursue our traditional emphasis on the export of primary produce.

While there is a strong argument for trying to become at least more self-sufficient in value-added goods and services, we might not want to emulate Japanese society for a variety of reasons. Not only is there a dedication to the work ethic, and a degree of pressure to conform that we would probably find uncomfortable, but the Japanese approach suffers from a fundamental problem of ecological illiteracy. When he visited Australia in 1988, the Canadian scientist Dr David Suzuki urged us not to follow the path of Japan, the country of his ancestors. He pointed out that we are self-sufficient in most important ways, whereas Japanese society can only survive with massive imports of food, minerals and energy. In terms of political and economic independence, there are obviously great advantages in having a lifestyle which is both sustainable and self-sufficient.

Let me summarize what I have said in this [article]. I have discussed the complex economic role of technology in modern industrial society. The future is likely to see technology playing an increasingly important role, not just in economic terms but also as a social and political force. The world of today is one in which the economic and political roles of nations are increasingly determined by the way they manage technological change. This poses a clear challenge for Australia. The recently retired professor of science and society at Bradford University, Tom Stonier, has said that technological change represents opportunity to those who are well educated and aware, but is a threat to others. That observation applies to nations as well as to individuals.

[. . .]

1.6

No Growth in a Finite World

S. Irvine

[Compared with other articles in the Reader, this is particularly trenchant; the author's values about technology and economic growth are made very clear. This is one of the main reasons for its selection. It highlights in a radical way some of the issues being created by the presently dominant economic model of technological development and continuing economic growth for all. Some explanatory footnotes are given on the term 'economic growth' and on some of the documents mentioned in the article.]

Many people now agree that society cannot continue on its present course without committing ecological suicide. This approaching crisis has created a new orthodoxy. Among turquoise Tories and greened Socialists – as well as in the proliferating pressure groups and non-governmental organizations – there is a consensus that salvation lies in something variously called 'sustainable', 'balanced', 'controlled', 'differentiated', 'qualitative', 'green' or even simply 'good' growth and development. On this bandwagon, there is no place for the anti-growth perspectives expressed, in the early 1970s, by the Club of Rome's *Limits to Growth* study[1] and by *The Ecologist's Blueprint for Survival*. A recent Fabian Society pamphlet, *Sustainable Development* by Michael Jacobs, reflects this new consensus when it argues that 'the concept of overall zero growth is meaningless', dismissing the debate about the 'desirability of economic growth'[2] as 'not a helpful one'. 'In short,' Jacobs has argued elsewhere, 'it makes no sense to be for [economic growth] or against it'.

The analysis and policies contained in the Brundtland Report, *Our Common Future*,[3] have encapsulated the essence of born-again growth strategies. 'Green growthism' even has its own coffee-table book (*The Gaia Atlas of Planet Management*) and its guide to sustainable shopping (*The Green Consumer Guide*).

The Brundtland Report exemplifies all that is wrong with green growthism. It is basically about sustaining business-as-usual, rather than bringing about a fundamental revolution in values, lifestyles and social structures. The report calls for a 'new era of economic growth', and argues that this is the way to reduce global poverty and 'create the capacity we need to overcome the environmental crisis'. It proposes a target of some 3–4 per cent 'new growth' for countries like Britain, and higher figures for less developed ones. It advocates 'an expanding world economy' and talks of 'a five- to tenfold increase in manufacturing output'.

The door to sustainable growth is to be found in 'less material- and energy-intensive activities' and improved efficiency. The bad old days of ungreen growth will be left behind. Instead, 'policy-makers will ensure that growing economies remain firmly attached to their ecological roots and that those roots are protected and nurtured so that they may support growth over the long term'.

The Brundtland Report's ultimate goal would appear to be western lifestyles for all. The report, for example, calls for increases in manufacturing output that 'raise developing-world consumption to industrialized-world levels by the time population growth rates level off next century'.

However, the word 'common', in *Our Common Future*, would appear to include only human members of the biosphere. Gro Harlem Brundtland's foreword to the report defines human well-being as the 'ultimate goal'. If need be, other life forms will be sacrificed for that purpose. Greater use of pesticides and chemical fertilizer and more ranching projects are, for example, envisaged; the consequence of this can only be less non-human life. If the earth has friends like this, it scarcely needs enemies.

If Dr Johnson were alive today, he might rephrase his famous dictum to: 'Sustainability is the last refuge of the scoundrel'. Green growthism uses the language of the ecological movement to dress up the same old goals. Words like 'clean', 'renewable' and 'recycled' pop up everywhere.

Michael Jacobs's pamphlet on sustainable development, for example, claims that 'natural-fibre clothes' (unlike, presumably, synthetic ones) have 'relatively little environmental impact'. In fact, large-scale cotton monoculture and sheep grazing produce soil erosion and other impacts as unsustainable as the resource depletion and pollution of oil production and processing (to which cotton and wool output remains linked via the use of petrol used in machinery, artificial fertilizers and various pesticides).

The renewability of a resource base does not of itself make it appropriate. Large-scale hydroelectric dams, for example, can be non-polluting and, judged in isolation, renewable: yet they cause massive environmental degradation. Fuel from crops is yet another illusory free lunch.

The claim that the new growth will be cleaner also depends on the efficacy of clever gadgetry fitted to polluting processes. The greenhouse effect provides some revealing counter-examples. The only practicable way to prevent the release of carbon is to burn less wood and less coal. The only way to control the build-up of another greenhouse gas, methane, is to limit the amount of cattle ranching and of rice growing. In fact, technological gadgets merely shift the problem around, often at the expense of more energy and material inputs – and therefore more pollution. Catalytic converters for cars, for example, cost money and energy, generate new pollutants and fail to solve more serious problems such as carbon-dioxide emissions.

Parallel deceptions are found in the demand for 'socially useful' production. All forms of production put pressure on ecological systems. Nature's economy keeps the same accounts for the energy and raw materials embodied in ambulances or armoured cars.

Arguments among theologians of green growthism about the meaning of 'growth' and 'development' have shrouded the real issue: the total *throughput* of physical space, energy and raw materials in the human economy. The much maligned limits-to-growth perspective is, in fact, the key to any understanding of why humanity faces its present dangers and why only a rejection of the whole growth paradigm will guide us to a more sustainable future.

Green growthism, by contrast, still regards 'wealth' as the cure for all ills, including global poverty and population growth. The language of 'moreness' excludes that of 'enoughness'. Mother Earth is still treated as an inefficient and, indeed, disorderly bitch to be tamed, domesticated and manipulated in the service of greater human consumption.

The starting point of the 'limits' model is the recognition that there is an optimal size for every structure, be it a plant, animal, machine, institution, social system or eco-system. As ecologist Eugene Odum once put it: 'Growth beyond the optimum becomes cancer'. Contrary to the claim, made by Michael Jacobs, that 'sustainability does not rule out growth', Odum and many others have shown that youthful growth does, in fact, give way to stability and order in sustainable systems. The expansion of any part of a system beyond its optimum is self-defeating, since it can only displace other parts: thereby it first disrupts, then destroys, the overall system, itself included. This is precisely what human society is now doing to the ecological envelope.

From this basis, what will be called the limits-to-throughput model (LTM) makes two deductions: first, that the environmental crisis is the direct and inevitable consequence of excessive global consumption of energy and raw materials; and second that, taken as a whole, global society is in a general state of overdevelopment.

Contrary to the popular, quite mistaken, notion that 'ecology and economy are inter-dependent', the human-constructed economy and the technologies that serve it are totally dependent on the economy of nature, whose ecological systems represent the biophysical foundation of all wealth. They include, of course, direct benefits in the form of specific resources such as foodstuffs, medicines, natural insecticides, clothing materials, wood, biomass fuels, oils, rubber, glues and many more.

However, far more important, and vital to the very habitability of the planet, are services of a more indirect character. They include: the capture, conversion and storage of solar energy income; the maintenance of atmospheric balances; air purification; the amelioration of weather; the provision of shelter; the signalling of pollution problems; the regulation of the water cycle; the disposal of wastes and recycling of nutrients; the generation and maintenance of soil fertility; pest and disease control; pollination; the maintenance of the 'library' of genetic information vital for new medicines, foodstuffs and other needs; the human pleasure, inspiration and inner renewal derived from contact with unmodified nature.

There are intrinsic limits to the resources humans can take from environmental systems on a sustainable basis. The transformation of energy and raw materials in the human economy generates wastes, which are not 'thrown away' but return to the 'sink' of air, soil and water. There are equivalent limits to what can be sustainably reassimilated by the environment. As a result, nature's economy, the ultimate 'means of production', can bake only a limited cake.

There are three main biological and physical reasons why this must be the case. The most obvious limit is the earth itself, which is finite not only in size and resources available, but also in its capacity to absorb the waste by-products thus generated. This would not be so significant if it were not for limit two: the second law of thermodynamics, or entropy. Basic physical laws tell us that all energy usage is, of course, a one-way process. You cannot relight a fire from yesterday's ashes, nor can a car run on exhaust fumes. Further, at every stage in the extraction, processing and usage of a raw material, some of it is lost for ever.

With raw material, unlike energy, there is the option of recycling. However, recovery and restoration to a usable form takes more energy and creates more pollution. Recycling cannot, therefore, underwrite sustainably a growing throughput in the economy. The expansion of what might be called the 'technosphere' is inevitably at the expense of the ecosphere, on whose integrity all human societies, no matter how sophisticated, remain dependent.

The original limits-to-growth debate in the early 1970s focused primarily on the question of resource availability. Though we will soon face a general crisis, a shortage of resources is, in fact, the least of our problems. It is the *side-effects* from increased use of energy and material resources – pollution – on life-supporting environmental systems, and the consequences of their impairment, that constitute the most fundamental outer limit to growth.

These are, however, still not the only limits. The LTM model recognizes constraints within society itself. The bigger and more complicated an organization, for example, the more it overtaxes human capacity to predict, plan and manage.

Many people recognize that the world's economic cake is badly shared out. Green growthism argues that we need more economic production to provide the clean water supplies, good housing, and many other facilities the poor lack. Their suffering, however, is not due to a lack of growth. It is due to inequality and exploitation. Their position has worsened, not improved, in countries such as Brazil, where economic expansion has been given priority.

If all citizens in the so-called Third World could have the same standard of living as a citizen in the industrialized world, total global output would need to be more than 130 times what it was in 1979. The only feasible alternative is a radical redistribution of the world's resources: in one Canadian estimate, people there would get 20 per cent of the energy they currently consume. Although organizations such as the Club of Rome were attacked for élitism, the implications of the limit-to-growth perspective are profoundly egalitarian.

The final card in the green-growth pack claims that we need more growth to defuse the population time bomb. In a completely ahistorical comparison with the stabilization of population after the Industrial Revolution in Europe, it is argued that more production is necessary to raise living standards so that this happens elsewhere. However, as we saw above, the physical resources are simply not available to give everyone the level of per-capita consumption of countries where population growth has levelled off.

Moreover, poverty is only one of many reasons why adults have large numbers of offspring. Large families are not unknown among the rich: some of the fastest-rising populations are in countries, such as Kenya, with significant economic development. Indeed, there is a good case for arguing that the social disruption caused by the pursuit of economic growth in the Third World created the conditions for the population explosion. Conversely, many traditional and supposedly 'poor' societies managed to regulate their numbers, not because of high levels of infant mortality but by social conventions.

In a finite, entropy-bound and ecologically interconnected system, sustaining more of one thing must mean less of something else. This is seen at its most dramatic in the expansion of the human species at the expense of non-human life forms.

There is also a trade-off between the affluence of the world's rich ghettos and the grinding poverty outside, especially in Africa and southern Asia. In the past 50 years, for one example, the US has consumed more fossil fuels and minerals than the rest of humanity has done in the whole of recorded history. The increased 'throughput' of goats and sheep, for another, has had disastrous effects in the 'wet deserts' of the British uplands, in the bare hills all around the Mediterranean basin, and in the dry plains of Africa.

Green growthism pins the blame for our problems firmly on a 'rogue' element in the present mix of production processes, often chosen by profiteering capitalists or blundering bureaucrats. On closer examination, many of the allegedly cleaner and more renewable alternatives turn out to be far from satisfactory solutions. Electric cars, for instance, are at one level non-polluting: but this ignores the resources consumed, and pollution caused, by electricity generation; the energy and raw materials to make the car and the batteries; and the roads, traffic lights and garages they still need.

Others pin their hopes on the so-called sunrise industries and on 'advanced' materials. History is repeating itself today with 'advanced materials', many of which emit in their manufacture effluents similar to those from more traditional industry.

However, the most dangerous forms of green growthism's pursuit of cleaner and more efficient growth are genetic engineering and biotechnology. Their safety depends on flawless design and construction, no malfunctions, no operating errors and perfect protection from acts of God and people. They seem likely to be the first step to pollution not just of the environment, but of the evolutionary process itself.

Green growth is therefore no different from any other form of expansion. It, too, will face the barrier of increasingly negative trade-offs and insurmountable limiting factors. Savings on waste and built-in obsolescence may provide the wherewithal for a redirection of existing resources. But further expansion can be attained only by putting even more of the earth under concrete, under the plough and the open-cast excavator, or beneath impounded water.

Limits only become meaningful once there is a danger of transgressing them. Many past societies have done so and paid the price. These cases of eco-suicide were localized; today, the symptoms are everywhere. Overdevelopment is the best word to describe this situation: it is a combination of overpopulation and overconsumption. This is not only unsustainable, but also unfair. The greater the resources consumed today, the less there can be for future generations and other species.

In fact, the human predicament is that of someone suffering from obesity: to get better we have to slim down. Clever gadgetry might help 'fine-tune' the engine but does not remove the need to make it 'rev' at a lower overall level.

Writing of a 'sustainable Earth society', Australian physicist G. Taylor Miller reminds us of other species. Once provision has been made for their right to a share of the earth's resources and living space, we can begin to think of sharing out fairly among the world's human population. This would be done by taking from the environment only resources that nature can renew for the sake of future generations.

A deeper and more holistic approach would be to think of a society based on the sustainable and balanced satisfaction of different kinds of present and future human needs — material, psychological and spiritual — and those of other species. It would oppose those values, lifestyles and institutions that try to satisfy the aspirations of one community, culture, generation or species by destroying another.

Such a programme requires a drastic change — though not one bigger than that which accompanied the Industrial Revolution — in social values, our economic system and the choice of technologies we use. As the Rolling Stones put it, 'you can't always get what you want'. Humanity will have to learn to live with the limited income it can sustainably take from nature's capital.

Notes

(1)   *Limits to Growth* is the book resulting from the technological forecasting studies of a group of researchers at the Massachusetts Institute of Technology (MIT) led by Professor Dennis Meadows. It was published in 1972. They used a complex 'world

model' to forecast future trends in resource depletion, population and pollution. One of their main conclusions was:

> 'If the present trends in world population, industrialization, pollution, food production and resource depletion continue unchanged, the limits to growth on this planet will be reached sometime within the next one hundred years. The most probable result will be a rather sudden and uncontrollable decline in population and industrial capacity.'

(2) In national terms, *economic growth* is the increase in production which contributes to an increase in national income. Conventionally, it is measured in terms of the *gross domestic product* (GDP) which is a financial estimate of the total flow of goods and services produced by the economy over a specified period – usually a year. (This is distinct from the *gross national product* (GNP) which is GDP plus income from overseas investment and minus domestic income earned by people living abroad.)

Economic growth is normally measured as a *percentage* over a given period of time (usually a year). One is aware of this growth rate being used as an indicator of the state of the economy – in comparison with that of our 'competitors' or of previous governments. Anything which grows by a percentage in a series of time periods grows *exponentially*. This is what compound interest does.

The important characteristic of an exponential (compound interest) curve is its *doubling time*. A useful guide is that the doubling time is given by 70 divided by the rate of interest. Even an economic growth rate of 2% – conventionally regarded as modest – means a quadrupling within a person's expected lifetime. An important implication of this is that, given very little recycling or similar conservation processes, consumption of resources and energy, and the production of pollutants will follow the same exponential trend.

(3) The Brundtland report *Our Common Future* is the result of an urgent call, in December 1983, by the General Assembly of the United Nations to the World Commission on Environment and Development to formulate a 'global agenda for change'.

The key point of the report is that it envisages that economic growth can continue by means of *sustainable development*. Sustainable development is defined as a pattern of activity which meets the needs of the present generation without reducing the options available to future generations. Thus, a technological activity must not deplete natural resources or make an unacceptable impact on the natural environment. (It should be added that social stability, local, national or international, should not be threatened either.)

1.7

Feminist Critiques of Science and Technology

J. Wajcman

■ From science to technology

While there has been a growing interest in the relationship of science to society over the last decade, there has been an even greater preoccupation with the relationship between technology and social change. Debate has raged over whether the 'white heat of technology' is radically transforming society and delivering us into a post-industrial age. A major concern of feminists has been the impact of new technology on women's lives, particularly on women's work. The introduction of word processors into the office provided the focus for much early research. The recognition that housework was also work, albeit unpaid, led to studies on how the increasing use of domestic technology in the home affected the time spent on housework. The exploitation of Third World women as a source of cheap labour for the manufacture of computer components has also been scrutinized. Most recently there has been a vigorous debate over developments in reproductive technology and the implications for women's control over their fertility.

Throughout these debates there has been a tension between the view that technology would liberate women – from unwanted pregnancy, from housework and from routine paid work – and the obverse view that most new technologies are destructive and oppressive to women. For example, in the early seventies, Shulamith Firestone (1970) elaborated the view that developments in birth technology held the key to women's liberation through removing from them the burden of biological motherhood. Nowadays there is much more concern with the negative implications of the new technologies, ironically most clearly reflected in the highly charged debate over the new reproductive technologies.

A key issue here is whether the problem lies in men's domination of technology, or whether the technology is in some sense inherently patriarchal. If women were in control, would they apply technology to more benign ends?

In the following discussion on gender and technology, I will explore these and related questions.

An initial difficulty in considering the feminist commentary on technology arises from its failure to distinguish between science and technology. Feminist writing on science has often construed science purely as a form of knowledge, and this assumption has been carried over into much of the feminist writing on technology. However, just as science includes practices and institutions, as well as knowledge, so too does technology. Indeed, it is even more clearly the case with technology because technology is primarily about the creation of artefacts. This points to the need for a different theoretical approach to the analysis of the gender relations of technology, from that being developed around science.

Perhaps this conflation of technology with science is not surprising given that the sociology of scientific knowledge over the last ten years has contested the idea of a non-controversial distinction between science and technology. John Staudenmaier (1985, pp. 83–120) comments that although the relationship between science and technology has been a major theme in science and technology studies, the discussion has been plagued by a welter of conflicting definitions of the two basic terms. The only consensus to have emerged is that the way in which the boundaries between science and technology are demarcated, and how they are related to each other, change from one historical period to another.

In recent years, however, there has been a major re-orientation of thinking about the form of the relationship between science and technology. The model of the science–technology relationship which enjoyed widespread acceptance over a long period was the traditional hierarchical model which treats technology as applied science. This view that science discovers and technology applies this knowledge in a routine uncreative way is now in steep decline. 'One thing which practically any modern study of technological innovation suffices to show is that far from applying, and hence depending upon, the culture of natural science, technologists possess their own distinct cultural resources, which provide the principal basis for their innovative activity' (Barnes and Edge, 1982, p. 149). Technologists build on, modify and extend existing technology but they do this by a creative and imaginative process. And part of the received culture technologists inherit in the course of solving their practical problems is non-verbal; nor can it be conveyed adequately by the written word. Instead it is the individual practitioner who transfers practical knowledge and competence to another. In short, the current model of the science–technology relationship characterizes science and technology as distinguishable sub-cultures in an interactive symmetrical relationship.

Leaving aside the relationship between technology and science, it is most important to recognize that the word 'technology' has at least three different layers of meaning. Firstly, 'technology' is a form of knowledge, as Staudenmaier emphasizes.[1] Technological 'things' are meaningless without the 'know-how' to use them, repair them, design them and make them. That know-how often cannot be captured in words. It is visual, even tactile, rather than simply verbal or mathematical. But it can also be systematized and taught, as in the various disciplines of engineering.

Few authors however would be content with this definition of technology as a form of knowledge. 'Technology' also refers to what people do as well as what they know. An object such as a car or a vacuum cleaner is a technology, rather than an arbitrary lump of matter, because it forms part of a set of human activities. A computer without programs and programmers is simply a useless collection of bits of metal, plastic and silicon. 'Steel-making', say, is a technology: but this implies that the technology includes what steelworkers do, as well as the furnaces they use. So 'technology' refers to human activities and practices. And finally, at the most basic level, there is the 'hardware' definition of technology, in which it refers to sets of physical objects, for example, cars, lathes, vacuum cleaners and computers.

These different layers of meaning of 'technology' are worth bearing in mind in what follows.

The rest of this chapter will review the theoretical literature on gender and technology, which in many cases mirrors the debates about science outlined above. However, feminist perspectives on technology are more recent and much less theoretically developed than those which have been articulated in relation to science. One clear indication of this is the preponderance of edited collections which have been published in this area.[2] As with many such collections, the articles do not share a consistent approach or cover the field in a comprehensive fashion. Therefore I will be drawing out strands of argument from this literature rather than presenting the material as coherent positions in a debate.

■ Hidden from history

To start with, feminists have pointed out the dearth of material on women and technology, especially given the burgeoning scholarship in the field of technology studies. Even the most perceptive and humanistic works on the relationship between technology, culture and society rarely mention gender. Women's contributions have by and large been left out of technological history. Contributions to *Technology and Culture*, the leading journal of the history of technology, provide one accurate barometer of this. Joan Rothschild's (1983, pp. xii–xiv) survey of the journal for articles on the subject of women found only four in 24 years of publishing. In a more recent book about the journal, Staudenmaier (1985, p. 180) also notes the extraordinary bias in the journal towards male figures and the striking absence of a women's perspective. The history of technology represents the prototype inventor as male. So, as in the history of science, an initial task of feminists has been to uncover and recover the women hidden from history who have contributed to technological developments.

There is now evidence that during the industrial era, women invented or contributed to the invention of such crucial machines as the cotton gin, the sewing machine, the small electric motor, the McCormick reaper, and the Jacquard loom (Stanley, in press). This sort of historical scholarship often relies heavily on patent records to recover women's forgotten inventions. It has been noted that many women's inventions have been credited to their husbands because they actually appear in patent records in their husbands' name. This is explained in terms of women's limited property rights, as well as the general ridicule afforded women inventors at that time (Pursell, 1981; Amram, 1984; Griffiths, 1985). Interestingly, it may be that even the recovery of women inventors from patent records seriously underestimates their contribution to technological development. In a recent article on the role of patents, Christine MacLeod (1987) observes that prior to 1700 patents were not primarily about the recording of the actual inventor, but were instead sought in the name of financial backers.[3] Given this, it is even less surprising that so few women's names are to be found in patent records.

For all but a few exceptional women, creativity alone was not sufficient. In order to participate in the inventive activity of the Industrial Revolution, capital as well as ideas were necessary. It was only in 1882 that the Married Women's Property Act gave English women legal possession and control of any personal property independently of their husbands. Dot Griffiths (1985) argues that the effect of this was to virtually exclude women from participation in the world of the inventor–entrepreneur. At the same time, women were being denied access to education and specifically to the theoretical grounding in mathematics and mechanics upon which so many of the inventions and innovations of the period were based. As business activities expanded and were moved out of the home, middle-class women were increasingly left to a life of enforced leisure. Soon the appropriate education for girls became 'accomplishments' such as embroidery and music – accomplishments hardly conducive to participation in the world of the inventor–entrepreneur. In the current period, there has been considerable interest in the possible contributions which Ada Lady Lovelace, Grace Hopper and other women may have made to the development of computing. Recent histories of computer programming provide substantial evidence for the view that women played a major part.[4]

To fully comprehend women's contributions to technological development, however, a more radical approach may be necessary. For a start, the traditional conception of technology too readily defines technology in terms of male activities. As I have pointed out above, the concept of technology is itself subject to historical change, and different epochs and cultures had different names for what we now think of as technology. A greater emphasis on women's activities immediately suggests that females, and in particular black women, were among the first technologists. After all, women were the main gatherers, processors and storers of plant food from earliest human times onward. It was therefore logical that they should be the ones to have invented the tools and methods involved in this work such as the digging stick, the carrying sling, the reaping knife and sickle, pestles and pounders. In this vein, Autumn Stanley (in press) illustrates women's early achievements in horticulture and agriculture, such as the hoe, the scratch plow, grafting, hand pollination, and early irrigation.

If it were not for the male bias in most technology research, the significance of these inventions would be acknowledged. As Ruth Schwartz Cowan notes:

> 'The indices to the standard histories of technology . . . do not contain a single reference, for example, to such a significant cultural artifact as the baby bottle. Here is a simple implement . . . which has transformed a fundamental human experience for vast numbers of infants and mothers, and been one of the more controversial exports of Western technology to underdeveloped countries − yet it finds no place in our histories of technology.' (1979, p. 52)

There is important work to be done not only in identifying women inventors, but also in discovering the origins and paths of development of 'women's sphere' technologies that seem often to have been considered beneath notice.

■ A technology based on women's values?

During the 1980s, feminists have begun to focus on the gendered character of technology itself. Rather than asking how women could be more equitably treated within and by a neutral technology, many feminists now argue that Western technology itself embodies patriarchal values. This parallels the way in which the feminist critique of science evolved from asking the 'woman question' in science to asking the more radical 'science question' in feminism. Technology, like science, is seen as deeply implicated in the masculine project of the domination and control of women and nature.[5] Just as many feminists have argued for a science based on women's values, so too has there been a call for a technology based on women's values. In Joan Rothschild's (1983) preface to a collection on feminist perspectives on technology, she says that: 'Feminist analysis has sought to show how the subjective, intuitive, and irrational can and do play a key role in our science and technology'. Interestingly, she cites an important male figure in the field, Lewis Mumford, to support her case. Mumford's linking of subjective impulses, life-generating forces and a female principle is consistent with such a feminist analysis, as is his endorsement of a more holistic view of culture and technological developments.

Other male authors have also advocated a technology based on women's values. Mike Cooley is a well-known critic of the current design of technological systems and he has done much to popularize the idea of human-centred technologies. In *Architect or Bee?* (1991, p. 88) he argues that technological change has 'male values' built into it: 'the values of the White Male Warrior, admired for his strength and speed in eliminating the weak, conquering competitors and ruling over vast armies of men who obey his every instruction . . . Technological change is starved of the so-called female values such as intuition, subjectivity, tenacity and compassion'. Cooley sees it as imperative that more women become involved in science and technology to challenge and counteract the built-in male values: that we cease placing the objective above the subjective, the rational above the tacit, and the digital above analogical representation. In *The Culture of Technology*, Arnold Pacey (1983) devotes an entire chapter to 'Women and Wider Values'. He outlines three contrasting sets of values involved in the practice of technology — firstly, those stressing virtuosity, secondly, economic values and thirdly, user- or need-oriented values. Women exemplify this third 'responsible' orientation, according to Pacey, as they work with nature in contrast to the male interest in construction and the conquest of nature.

Ironically the approach of these male authors is in some respects rather similar to the eco-feminism that became popular amongst feminists in the 1980s. This marriage of ecology and feminism rests on the 'female principle', the notion that women are closer to nature than men and that the technologies men have created are based on the domination of nature in the same way that they seek to dominate women. Eco-feminists concentrated on military technology and the ecological effects of other modern technologies. According to them, these technologies are products of a patriarchal culture that 'speaks violence at every level' (Rothschild, 1983, p. 126). An early slogan of the feminist anti-militarist movement, 'Take the Toys from the Boys', drew attention to the phallic symbolism in the shape of missiles. However, an inevitable corollary of this stance seemed to be the representation of women as inherently nurturing and pacifist. The problems with this position have been outlined above in relation to science based on women's essential values. We need to ask how women became associated with these values. The answer involves examining the way in which the traditional division of labour between women and men has generally restricted women to a narrow range of experience concerned primarily with the private world of the home and family.

Nevertheless, the strength of these arguments is that they go beyond the usual conception of the problem as being women's exclusion from the processes of innovation and from the acquisition of technical skills. Feminists have pointed to all sorts of barriers — in social attitudes, girls' education and the employment policies of firms — to account for the imbalance in the number of women in engineering. But rarely has the problem been identified as the way engineering has been conceived and taught. In particular, the failure of liberal and equal opportunity policies has led authors such as Cynthia Cockburn (1985) to ask whether women actively resist entering technology. Why have the women's training initiatives designed to break men's monopoly of the building trades, engineering and information technology not been more successful? Although schemes to channel women into technical trades have been small-scale, it is hard to escape the conclusion that women's response has been tentative and perhaps ambivalent.

I share Cockburn's view that this reluctance 'to enter' is to do with the sex-stereotyped definition of technology as an activity appropriate for men. As with science, the very language of technology, its symbolism, is masculine. It is not simply a question of acquiring skills, because these skills are embedded in a culture of masculinity that is largely coterminous with the culture of technology. Both at school and in the workplace this culture is incompatible with femininity. Therefore, to enter this world, to learn its language, women have first to forsake their femininity.

■ Technology and the division of labour

I will now turn to a more historical and sociological approach to the analysis of gender and technology. This approach has built on some theoretical foundations provided by contributors to the labour process debate of the 1970s. Just as the radical science movement had sought to expose the class character of science, these writers attempted to extend the class analysis to technology. In doing so, they were countering the theory of 'technological determinism' that remains so widespread.

According to this account, changes in technology are the most important cause of social change. Technologies themselves are neutral and impinge on society from the outside; the scientists and technicians who produce new technologies are seen to be independent of their social location and above sectional interests. Labour process analysts were especially critical of a technicist version of Marxism in which the development of technology and productivity is seen as the motor force of history. This interpretation represented technology itself as beyond class struggle.

With the publication of Harry Braverman's *Labor and Monopoly Capital* (1974), there was a revival of interest in Marx's contribution to the study of technology, particularly in relation to work. Braverman restored Marx's critique of technology and the division of labour to the centre of his analysis of the process of capitalist development. The basic argument of the labour process literature which developed was that capitalist–worker relations are a major factor affecting the technology of production within capitalism. Historical case studies of the evolution and introduction of particular technologies documented the way in which they were deliberately designed to deskill and eliminate human labour.[6] Rather than technical inventions developing inexorably, machinery was used by the owners and managers of capital as an important weapon in the battle for control over production. So, like science, technology was understood to be the result of capitalist social relations.

This analysis provided a timely challenge to the notion of technological determinism and, in its focus on the capitalist division of labour, it paved the way for the development of a more sophisticated analysis of gender relations and technology. However, the labour process approach was gender-blind because it interpreted the social relations of technology in exclusively class terms. Yet, as has been well established by the socialist feminist current in this debate, the relations of production are constructed as much out of gender divisions as class divisions. Recent writings (Cockburn, 1983, 1985; Faulkner and Arnold, 1985; McNeil, 1987) in this historical vein see women's exclusion from technology as a consequence of the gender division of labour and the male domination of skilled trades that developed under capitalism. In fact, some argue that prior to the industrial revolution women had more opportunities to acquire technical skills, and that capitalist technology has become more masculine than previous technologies.

I have already described how, in the early phases of industrialization, women were denied access to ownership of capital and access to education. Shifting the focus, these authors show that the rigid pattern of gender divisions which developed within the working-class in the context of the new industries laid the foundation for the male dominance of technology. It was during this period that manufacturing moved into factories, and home became separated from paid work. The advent of powered machinery fundamentally challenged traditional craft skills because tools were literally taken out of the hands of workers and combined into machines. But as it had been men who on the whole had technical skills in the period before the Industrial Revolution, they were in a unique position to maintain a monopoly over the new skills created by the introduction of machines.

Male craft workers could not prevent employers from drawing women into the new spheres of production. So instead they organized to retain certain rights over technology by actively resisting the entry of women to their trades. Women who became industrial labourers found themselves working in what were considered to be unskilled jobs for the lowest pay. 'It is the most damning indictment of skilled working-class men and their unions that they excluded women from membership and prevented them gaining competences that could have secured them a decent living' (Cockburn, 1985, p. 39). This gender division of labour within the factory meant that the machinery was designed by men with men in mind, either by the capitalist inventor or by skilled craftsmen. Industrial technology from its origins thus reflects male power as well as capitalist domination.

The masculine culture of technology is fundamental to the way in which the gender division of labour is still being reproduced today. By securing control of key technologies, men are denying women the practical experience upon which inventiveness depends. I noted earlier the degree to which technical knowledge involves tacit, intuitive knowledge and 'learning by doing'. New technology typically emerges not from sudden flashes of inspiration but from existing technology, by a process of gradual modification to, and new combinations of, that existing technology. Innovation is to some extent an imaginative process, but that imagination lies largely in seeing ways in which existing devices can be improved, and in extending the scope of techniques successful in one area into new areas. Therefore giving women access to formal technical knowledge alone does not provide the resources necessary for invention. Experience of existing technology is a precondition for the invention of new technology.

The nature of women's inventions, like that of men's, is a function of time, place and resources. Segregated at work and primarily confined to the private sphere of the household, women's experience has been severely restricted and therefore so too has their inventiveness. An interesting illustration of this point lies in the fact that women who were employed in the munitions factories during the First World War are on record as having redesigned the weaponry they were making.[7] Thus, given the opportunity, women have demonstrated their inventive capacity in what now seems the most unlikely of contexts.

■ Missing: the gender dimension in the sociology of technology

The historical approach is an advance over essentialist positions which seek to base a new technology on women's innate values. Women's profound alienation from technology is accounted for in terms of the historical and cultural construction of technology as masculine. I believe that women's exclusion from, and rejection of, technology is made more explicable by an analysis of technology as a culture that expresses and consolidates relations amongst men. If technical competence is an integral part of masculine gender identity, why should women be expected to aspire to it?

Such an account of technology and gender relations, however, is still at a general level.[8] There are few cases where feminists have really got inside the 'black box' of technology to do detailed empirical research, as some of the most recent sociological literature has attempted. Over the last few years, a new sociology of technology has emerged which is studying the invention, development, stabilization and diffusion of specific artefacts.[9] It is evident from this research that technology is not simply the product of rational technical imperatives. Rather, political choices are embedded in the very design and selection of technology.

Technologies result from a series of specific decisions made by particular groups of people in particular places at particular times for their own purposes. As such, technologies bear the imprint of the people and social context in which they developed. David Noble (1984, p. xiii) expresses this point succinctly as follows: 'Because of its very concreteness, people tend to confront technology as an irreducible brute fact, a given, a first cause, rather than as hardened history, frozen fragments of human and social endeavor'. Technological change is a process subject to struggles for control by different groups. As such, the outcomes depend primarily on the distribution of power and resources within society.

There is now an extensive literature on the history of technology and the economics of technological innovation. Labour historians and sociologists have investigated the relationship between social change and the shaping of production processes in great detail and have also been concerned with the influence of technological form upon social relations. The sociological approach has moved away from studying the individual inventor and from the notion that technological innovation is a result of some inner technical logic. Rather, it attempts to show the effects of social relations on technology that range from fostering or inhibiting particular technologies, through influencing the choice between competing paths of technical development, to affecting the precise design characteristics of particular artefacts. Technological innovation now requires major investment and has become a collective, institutionalized process. The evolution of a technology is thus the function of a complex set of technical, social, economic, and political factors. An artefact may be looked on as the 'congealed outcome of a set of negotiations, compromises, conflicts, controversies and deals that were put together between opponents in rooms filled with smoke, lathes or computer terminals' (Law, 1987, p. 406).

Because social groups have different interests and resources, the development process brings out conflicts between different views of the technical requirements of the device. Accordingly, the stability and form of artefacts depends on the capacity and resources that the salient social groups can mobilize in the course of the development process. Thus, in the technology of production, economic and social class interests often lie behind the development and adoption of devices. In the case of military technology, the operation of bureaucratic and organizational interests of state decision-making will be identifiable. Growing attention is now being given to the extent to which the state sponsorship of military technology shapes civilian technology.

So far, however, little attention has been paid to the way in which technological objects may be shaped by the operation of gender interests. This blindness to gender issues is also indicative of a general problem with the methodology adopted by the new sociology of technology. Using a conventional notion of technology, these writers study the social groups which actively seek to influence the form and direction of technological design. What they overlook is the fact that the absence of influence from certain groups may also be significant. For them, women's absence from observable conflict does not indicate that gender interests are being mobilized. For a social theory of gender, however, the almost complete exclusion of women from the technological community points to the need to take account of the underlying structure of gender relations. Preferences for different technologies are shaped by a set of social arrangements that reflect men's power in the wider society. The process of technological development is socially structured and culturally patterned by various social interests that lie outside the immediate context of technological innovation.

More than ever before technological change impinges on every aspect of our public and private lives, from the artificially cultivated food that we eat to the increasingly sophisticated forms of communication we use. Yet, in common with the labour process debate, the sociology of technology has concentrated almost exclusively on the relations of paid production, focusing in particular on the early stages of product development. In doing so they have ignored the spheres of reproduction, consumption and the unpaid production that takes place in the home. By contrast, feminist analysis points us beyond the factory gates to see that technology is just as centrally involved in these spheres.

Inevitably perhaps, feminist work in this area has so far raised as many questions as it has answered. Is technology valued because it is associated with masculinity or is masculinity valued because of the association with technology? How do we avoid the tautology that 'technology is masculine because men do it'? Why is women's work undervalued? Is there such a thing as women's knowledge? Is it different from 'feminine intuition'? Can technology be reconstructed around women's interests? These are the questions that abstract analysis has so far failed to answer. The character of salient interests and social groups will differ depending on the particular empirical sites of technology being considered. Thus we need to look in more concrete and historical detail at how, in specific areas of work and personal life, gender relations influence the technological enterprise. [. . .] While it is the case that men dominate the scientific and technical institutions, it is perfectly plausible that there will come a time when women are more fully represented in these institutions without transforming the direction of technological development. To cite just one instance, women are increasingly being recruited into the American space–defence programme but we do not hear their voices protesting about its preoccupations. Nevertheless, gender relations are an integral constituent of the social organization of these institutions and their projects. It is impossible to divorce the gender relations which are expressed in, and shape technologies from, the wider social structures that create and maintain them. In developing a theory of the gendered character of technology, we are inevitably in danger of either adopting an essentialist position that sees technology as inherently patriarchal, or losing sight of the structure of gender relations through an overemphasis on the historical variability of the categories of 'women' and 'technology'.

[. . .]

Notes

(1) Staudenmaier (1985, pp. 103–20) outlines four characteristics of technological knowledge–scientific concepts, problematic data, engineering theory, and technological skill. [See Article 1.2]

(2) A good cross-section of this material can be found in Trescott (1979); Rothschild (1983); Faulkner and Arnold (1985); McNeil (1987); Kramarae (1988). McNeil's

book is particularly useful as it contains a comprehensive bibliography which is organized thematically.

(3)   MacLeod (1987) suggests that although George Ravenscroft is credited in the patent records with being the 'heroic' inventor of lead-crystal glass, he was rather the purchaser or financier of another's invention. This study alerts us to the danger of assuming that patent records have always represented the same thing.

(4)   For a biography of Lady Lovelace, which takes issue with the view of her as a major contributor to computer programming, see Stein (1985). However, both Kraft (1977) and more recently Giordano (1988) have documented the extensive participation of women in the development of computer programming.

(5)   Technology as the domination of nature is also a central theme in the work of critical theorists, such as Marcuse, for whom it is capitalist relations (rather than patriarchal relations) which are built into the very structure of technology. 'Not only the application of technology but technology itself is domination (of nature and men) – methodical, scientific, calculated, calculating control. Specific purposes and interests of domination are not foisted upon technology "subsequently" and from the outside; they enter the very construction of the technical apparatus' (Marcuse, 1968, pp. 223–4).

(6)   See Part Two of MacKenzie and Wajcman (1985) for a collection of these case studies.

(7)   Amram (1984) provides a selection of the patents granted to women during the First World War.

(8)   Cockburn's (1983, 1985) work is one important exception.

(9)   For an introduction to this literature, see MacKenzie and Wajcman (1985); Bijker, Hughes and Pinch (1987).

■ References

Amram, F. (1984) The innovative woman. *New Scientist*, 24 May, 10–12.

Barnes, B. and Edge, D. (eds.) (1982) *Science in Context: Readings in the Sociology of Science*. Milton Keynes: Open University Press.

Bijker, W., Hughes, T. and Pinch, T. (eds.) (1987) *The Social Construction of Technological Systems*. Cambridge, MA: MIT Press.

Braverman, H. (1974) *Labor and Monopoly Capital: The Degradation of Work in the Twentieth Century*. New York: Monthly Review Press.

Cockburn, C. (1983) *Brothers: Male Dominance and Technological Change*. London: Pluto Press.

Cockburn, C. (1985) *Machinery of Dominance: Women, Men and Technical Know-How*. London: Pluto Press.

Cooley, M. (1991) *Architect or Bee? The Price of Technology*. (2nd edition). London: The Hogarth Press.

Cowan, R.S. (1979) From Virginia Dare to Virginia Slims: women and technology in American life. *Technology and Culture*, 20 (1), 51–63.

Faulkner, W. and Arnold, E. (eds.) (1985) *Smothered by Invention: Technology in Women's Lives*. London: Pluto Press.

Firestone, S. (1970) *The Dialectic of Sex*. New York: William Morrow and Co.

Giordano, R. (1988) *The Social Context of Innovation: A Case History of the Development of COBOL Programming Language*. Columbia University, Department of History.

Griffiths, D. (1985) 'The exclusion of women from technology' in Faulkner and Arnold, *op. cit.*

Kraft, P. (1977) *Programmers and Managers: the Routinization of Computer Programming in the United States*. New York: Springer Verlag.

Kramarae, C. (ed.) (1988) *Technology and Women's Voices*. New York: Routledge & Kegan Paul.

Law, J. (1987) Review Article: The structure of sociotechnical engineering – a review of the new sociology of technology. *The Sociological Review*, 35, 404–25.

MacKenzie, D. and Wajcman, J. (eds.) (1985) *The Social Shaping of Technology*. Milton Keynes: Open University Press.

MacLeod, C. (1987) Accident or design? George Ravenscroft's patent and the invention of lead-crystal glass. *Technology and Culture*, 28 (4), 776–803.

McNeil, M. (ed.) (1987) *Gender and Expertise*. London: Free Association Books.

Marcuse, H. (1968) *Negations*. London: Allen Lane.

Noble, D. (1984) *Forces of Production: a Social History of Industrial Automation*. New York: Knopf.

Pacey, A. (1983) *The Culture of Technology*. Oxford: Basil Blackwell.

Pursell, C. (1981) 'Women inventors in America'. *Technology and Culture*, 22 (3), 545–9.

Rothschild, J. (ed.) (1983) *Machina Ex Dea: Feminist Perspectives on Technology*. New York: Pergamon Press.

Stanley, A. (in press) *Mothers of Invention*. Metuchen, New Jersey: Scarecrow Press.

Staudenmaier, J. (1985) *Technology's Storytellers*. Cambridge, MA: MIT Press.

Stein, D. (1985) *Ada: A Life and Legacy*. Cambridge, MA: MIT Press.

Trescott, M.M. (ed.) (1979) *Dynamos and Virgins Revisited: Women and Technological Change in History*. Metuchen, New Jersey: Scarecrow Press.

PART 2

Features of Technological Activity

2.1

Modelling

J. Sparkes

'Facts are always less than what really happens' (*Nadine Gordimer*)

■ Introduction

In this article, the term 'modelling' does not refer to clay modelling or to the parading of the latest fashions in a clothes show. The most appropriate definition of modelling is 'the creation of simplified versions of reality for a particular purpose'. It is something we all do all the time in our minds and in the way we talk about things – so much so that we scarcely give it a thought. The purpose is usually to try to make sense of what we perceive. In technological design, however, modelling has a key role to play, so it is important that the process is properly understood. So I'll begin with the one aspect of modelling that we are all familiar with.

Take the example of the model aircraft that you can buy in toy shops. If one asks oneself whether it is a good model, one has first to ask what purpose it is supposed to serve, or what aim it is supposed to achieve. The aim is presumably to give pleasure to children (of all ages). Obviously, it is not supposed to represent in miniature *everything* about the real thing; it only represents certain features of it. Some children like models which look very like real aircraft; some like models which just have wings and wheels and a propeller with which they can create their own imaginative games; some like their models actually to fly. The different models each pick out some simplification(s) of the real thing which might please children, and ignore all the other features that real aircraft possess. This 'simplification for a purpose' is essentially what modelling is all about; but it can take many forms (Figure 1).

Not all model aircraft are toys. Aircraft manufacturers make models of their designs for the purpose of testing them in wind-tunnels and assessing their flying capabilities. These models have to be really accurate in some respects, particularly as regards their shape. Of course these models also do not represent

(a) A model that looks like a real aircraft (from the extensive Airfix range).

(b) A model that can actually fly.

Figure 1   Two kinds of model aircraft.

everything about the aircraft: they probably don't include miniature engines, passenger seats, etc. But manufacturers also use full-sized models of an aircraft's interior for trying out different seating layouts, etc., or as part of a selling strategy to attract customers.

Thus the first key point about modelling, at least in technology, is being clear about the *purpose* it is supposed to serve. Different aims or purposes lead directly to quite different models of the same real thing. Physical models of this kind, which look like the real thing, are called 'iconic models' after the Greek *eikon*, meaning 'image'.

Some physical models are not miniature or even full-sized representations of the real thing. For example, scientists make models of molecules, consisting of coloured balls — which represent atoms — joined together with sticks or springs, which represent chemical bonds. The purpose of these models is to enable people to visualize the structure of molecules. Such models played a crucial role in the discovery of the double helix structure of the DNA molecule (Figure 2). These models may be a billion times *bigger* than the 'real' things.

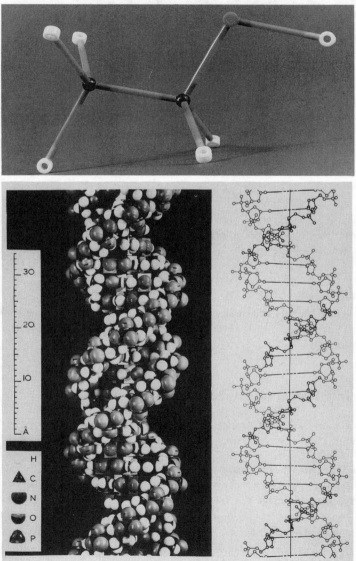

Figure 2 Ethanol molecule and double helix structure of DNA.

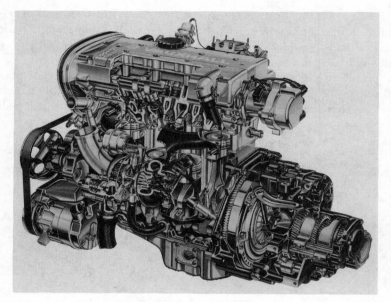

Figure 3   Cut-away engine.

Physical models, such as those already described, are widely used for educational purposes, especially working models which show how complex processes behave. For example:

- waves on water can be used to illustrate the dynamic characteristics (such as diffraction or reflection) of light waves;
- cut-away models of engines can be used to show the relative movements of all the various parts – the pistons, the valves, the crank-shaft and so on (Figure 3).

■ Other kinds of models

People use many kinds of models, other than iconic models, in their everyday lives for many different purposes. For example:

- An Ordnance Survey map is a model of the countryside, with contour lines used to represent height-above-sea-level and graphical symbols to represent features of the landscape.
- The well-known diagram of the London Underground system is a model. It is not scaled, so that, say, a distance of ten centimetres doesn't represent one kilometre; its purpose is to help travellers use the system effectively to get from place to place. It is almost useless as a guide to travel in any other way.

- A scale drawing of a bridge or car is a 'model' of its appearance every bit as much as is a toy aeroplane.

- A commonly used mathematical model is one which represents a car journey as a trip at a steady speed of, say, 60 miles per hour. We know this is not what the journey will actually be like, but it is a simplified model of it which enables us to calculate that a journey of 100 miles will take about 1 hour and 40 minutes.

If all these examples are to be thought of as models, a fairly broad definition of 'a model' is needed, such as:

- A model is a simplified or idealized version of reality created for a purpose

Like all brief definitions, this tends to embrace more than it should or exclude some examples that it shouldn't. But it includes the key ideas, and is therefore the definition of 'a model' that I shall be using in these notes. It oversteps the mark when it is stretched to embrace the works of art of painters and sculptors and even musical composers. We don't call such works of art 'models' even though they could be said to fit the definition. Thus the purposes included in the definition, at least where technological modelling is concerned, are usually practical ones, rather than aesthetic ones. So it is important not to try to stretch the definition too far. The key ideas it expresses are those of 'simplification or idealization', and 'purpose'. It is the purpose that determines the kind of simplification of reality that the model will represent. Similarly, the effectiveness of a model can only be judged in relation to its intended purpose(s). Criticizing a model because it doesn't fulfil some other purposes is not valid criticism. 'Modelling' is therefore the process of creating models which fulfil their intended purpose(s) successfully.

■ Models in science

One of the main activities in scientific research is the creation or discovery — depending on how you look at it — of explanatory models of phenomena in the real world. The scientific model of matter is that it consists of atoms. In other words, the idea that all matter could be broken down into tiny particles, with particular properties, was the 'model' of matter which replaced the 'continuum' models that preceded it. No one has seen an atom, it is too small; so all one can be sure about is that it is an imaginary creation, created for the purpose of explaining the properties of matter that we can all observe. By ascribing different properties to the different kinds of atoms it is possible to explain many of the phenomena that occur in the physical world. Thus *the purpose of scientific models is the explanation, and hence the rational prediction, of observed phenomena.*

But these different atomic properties themselves require explanation; so models of the structure of atoms have also been devised to provide these further explanations, consisting of even smaller particles, such as protons and electrons, as well as special kinds of forces, etc. The idea in science is to reduce as far as possible the number of explanatory concepts needed to explain everything!

The motivation for scientists to devise these models comes from the accompanying belief, which most scientists entertain, that their models are not models at all; they are descriptions of reality itself. The models are not 'created', they are 'discovered'. And scientists test their models so thoroughly against observations of real phenomena that they come to believe that they must be more than just models. They believe that atoms really exist as they have defined them in their models. But scientists have believed this about earlier models which have since been abandoned, so there may still be room for doubt, especially, for example about models of the universe – the Big Bang and so on. But, on the whole, modern scientific models are so successful in explaining physical phenomena – concerning inanimate matter at any rate – that it is difficult to see them as anything other than true descriptions of the physical world. That living matter remains mysterious, and that the behaviour of living creatures remains largely inexplicable at present is thought to be only a temporary state of affairs; before very long it will all be explicable. There are good grounds for doubting whether this will come about, but that is another matter.

Nevertheless scientific models tend to have a special place in technology, because they are an exceptionally reliable source of information on which to base technological designs. The three important points about scientific models are therefore:

(1) Their purpose is to explain, and hence to predict, observable phenomena.

(2) They are based on evidence which, in principle, can be obtained by anyone at any time. This is to ensure that subjective impressions or human bias are reduced to a minimum in the process of data gathering. But they cannot be deduced from evidence because the simplifications in modelling are not logical; they can only be inferred from it.

(3) Nevertheless scientific models have come to be widely regarded as true descriptions of reality itself, rather than as models of it.

■ Models in technology

In technology, models are mostly used as aids to design, though they are also used for selling or communication purposes. There are three main kinds of models: physical models, pictures and diagrams, and mathematical models:

☐ **Physical models**

Examples of these kinds of models have already been considered. They are physical models of particular aspects of a real product or system so that it can be tested or evaluated before the real 'thing' is produced. Other examples include:

- model cars for testing in wind tunnels, or models of boats for testing in water tanks, for 'drag' or streamlining (Figure 4);
- model buildings for judging their appearance relative to their environment, or for planning rooms, seating arrangements, etc., inside the buildings;
- models of an estuary, say, to test the effect of breakwaters, dams, etc., on silting, erosion, and so on.

Constructing miniature models of this kind is not simply a matter of making small-scale versions of the real things, because winds and water do not flow over small objects in the same way as they do over large objects. 'Scaling' is the term used to describe the business of making small models behave in the same way as full-sized ones; it is quite a science in itself.

☐ **Pictures and drawings**

Pictures, drawings and diagrams are of course widely used, not only in the first stages of design in order to try out design proposals, but also to communicate instructions to those who have to do the constructing. Examples include:

- Architects' drawings of buildings, either to show the overall layout of rooms, staircases, etc., or else as working drawings for the various building trades. Cross-sections, etc. represent what the building would look like if it were sliced through.
- Machine drawings for craftsmen to use in making mechanical parts (Figure 5). These diagrams often use simple conventions to indicate the three-dimensional character of the mechanical part, such as dashed lines to represent hidden edges, isometric projections, etc. There may have to be many thousands of drawings to represent all the parts of an aircraft or car.
- Circuit diagrams of electronic or electrical circuits. These diagrams are highly symbolic. Wires are drawn as straight lines with right-angle bends, and only show which components are connected to which; they do not represent the intended layout of the wires. Circuit elements, such as transistors, resistors, transformers, rectifiers, etc., are represented by standardized graphical symbols. Insulation is not represented at all; it is indicated simply by the white background of the paper.
- Maps also use symbols to represent the various features of the landscape, including the use of colour and contours to represent different altitudes or different kinds of ground cover.

Figure 4   Plane in wind tunnel.

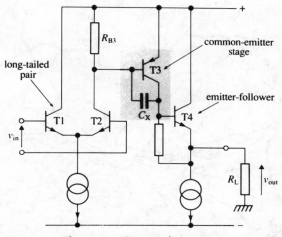

Figure 5 Circuit diagram.

- The well-known London Underground diagram (Figure 6) is a cross between a circuit diagram and a map!

Drawings and diagrams are obviously more abstract than physical models so people have to learn the conventions by means of which the drawings are related to the real things they represent.

☐ Mathematical models

Mathematical models are becoming more and more important as computers play an ever increasing role in technological design. So what are mathematical models?

A mathematical model is an equation or set of equations which represents – almost always in idealized form – the important properties of a real object or system, for the particular purposes of the model. The following are some examples.

- Consider again the 100 mile drive referred to earlier. Suppose it is from London to Birmingham. When you model the speed of the journey as '60 mph', you know perfectly well that your actual travelling speed will be different from this – sometimes faster, sometimes slower. But this model makes it easy to calculate your approximate travel time.*

---

* In creating a model like this people often say, 'let us assume that we travel at 60 mph'. Actually this is a misuse of the word 'assume', because they don't really *assume* any such thing; they know they are not going to travel at such a steady speed. What they really mean is, 'let us *suppose*, for the sake of a simple calculation, that we travel at 60 mph'. In other words they are creating a *simplified* model of their speed of travel for the purpose of arriving at a quick and approximate answer. *Assumptions* are possible true statements about reality, often standing in for unknown facts; *suppositions* are deliberate simplifications of reality usually for the purpose of making calculations easier, and so form a key part of modelling in technology.

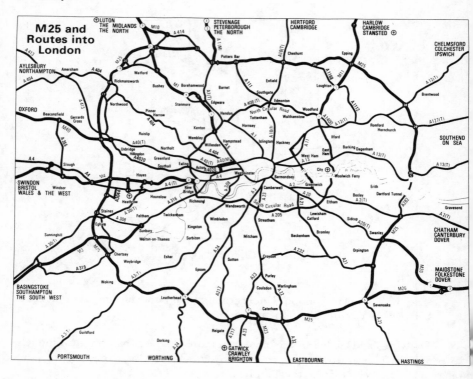

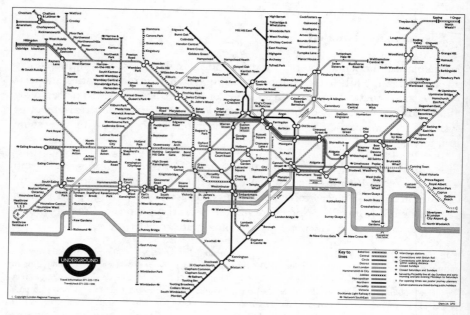

Figure 6   Map of London and the Underground 'map' (Registered user 92/1692).

The travel time is therefore $\dfrac{100\ \text{miles}}{60\ \text{mph}} \approx 1\dfrac{2}{3}$ hours

It is possible to create a more complicated and possibly more accurate model of the London–Birmingham journey (e.g. by supposing that the journey comprises 6 miles at 30 mph; 9 miles at 45 mph; 70 miles at 70 mph; 15 miles at 80 mph). This might give a more accurate representation of the journey, but the resulting journey time (1.59 hours) is more troublesome to calculate, and the 5 minutes' difference between the two results is hardly worth the effort.

Thus *an important aspect of modelling is judging how accurate to make one's model for the purpose in hand*. In practice, in technological design, this usually involves finding a model which fits reality as well as possible while at the same time giving rise to mathematical equations which can be solved sufficiently accurately. Very accurate models of reality all too often give rise to equations that are too difficult or too time-consuming to be solved economically, so the choice of appropriate simplifications often depends a good deal on successful previous experience.

- Physical science is much concerned with creating physical models of reality and then representing them mathematically. The behaviour of gases at different temperatures, for example, can be represented by Charles' Law as:

  For a given mass of gas at constant pressure,

  $$\dfrac{\text{its volume}}{\text{its temperature}} = \text{constant}$$

Although this 'law' of behaviour is not quite accurate, it is good enough for many purposes. For example, it enables you to calculate how much rise in temperature is needed to make a hot air balloon take off from the ground; or, together with Boyle's Law, it enables you to calculate that the pressure in your car tyres when they get hot on a Mediterranean holiday won't be enough to burst them.

- Mathematical models are widely used with electric circuits. The models are derived from basic science concepts, including, for example:

  - Resistance; defined as $\dfrac{\text{volts}}{\text{current}}$

  - Capacitance; defined as $\dfrac{\text{apparent current through the capacitor}}{\text{rate of change of voltage across the capacitor}}$

These concepts are not initially models of anything real, they are simply scientific concepts invented to explain observations. Resistors, capacitors, etc., do not occur naturally – as do gases described by Boyle's Law.

Technology comes into the picture when real resistors and capacitors are made in the image of these concepts. Resistors, for example, are made from pieces of materials for which the resistance — as defined in the above equation — remains the same even when the current through them is changed (i.e. they obey Ohm's Law). The better the technology of manufacture, the more accurate becomes the model. Sometimes (e.g. at high frequencies) and in some circuits, the model turns out to be not accurate enough, in which case it may be necessary to represent resistors by a more complicated model in order to predict circuit performance more accurately.

Similarly, *capacitors* are designed so that the capacitance, as defined in the above capacitance equation, remains constant even as the rate of change of voltage across the capacitor changes.

The purpose of these models is of course to facilitate calculations on the behaviour of electronic circuits, such as deducing the voltage across a resistor when you know the current through it.

- Some of the most complicated mathematical models are those used in simulators such as flight simulators. Here the behaviour of a particular aircraft in response to the pilot's control actions is modelled by means of mathematical equations. The solutions to these equations are translated, in a computer, into movements of the simulator and into visual images inside it, so that pilots sitting in it can experience all the sensations and problems that a real aircraft would create for them.

- Computer-aided design is another area where very complex sets of mathematical models are used. All the straight lines and curves of a drawing are represented by equations which are programmed into the computer. By giving, for example the end-points of a straight line, or the centre, radius and end-points of a curve, the computer will 'draw' the proposed design on the monitor. Designers can then easily manipulate the design to meet a particular specification without having to re-draw every proposed modification. They can also store each proposal before modifying it so that it can be returned to if the modifications turn out to show no improvement.

The use of mathematical models in engineering design has grown dramatically as the power and reliability of computers and software have increased. They provide a very powerful means of indicating, before it is built, whether a proposed design will be successful, whether in electronic circuit design, in aircraft design, in the design of cars, bridges, buildings and so on.

■ Conclusion

This business of modelling 'reality' raises some rather fundamental questions about reality and how we perceive it. What do we really know about reality?

We see and hear and feel that 'reality' is 'somewhere our there' – although we are also part of it – but it is difficult to bridge the gap between our subjective impressions of it and our belief that it is 'really there'. We cannot *know* for certain about it because all our impressions of it are processed by our eyes, ears and nerves. So all we can be sure about is that we form mental models of what we believe reality to be like!

So the *process* of modelling is not peculiar to science, maths and technology. In general its purpose is to make sense of our personal experiences, so that we can predict fairly reliably what the future is likely to bring. In general people have very different ideas of what 'makes sense', so people's perceptions of reality differ very markedly; though since the rise of science the influence of the supernatural has declined a good deal in people's perceptions of reality. Models in science and technology have the same broad aim of making sense of reality, but they depend on a range of special techniques aimed at improving the reliability of the predictions based on them. In science predictions are essentially logical deductions from repeatable data and well-tested scientific models. In technology the models are often based on science, or other academic disciplines, but also have to take into account many other factors in order to ensure that the designs and processes based on them are successful. Both science and technology do their best to eliminate from their models and predictions any factors based on the supernatural, or on guess-work, or on intuitive judgements or even on trial-and-error, though trial-and-success is normally acceptable!

2.2

Engineering Design

T392 course team

[Giving ourselves a problem and arriving at a solution is something we do all the time – preparing a meal, making a journey, buying a house. Designing is another example of this. Although the design process is basically simple, some of the steps and interactions among them may be complex. Nowadays 'quality', at least in an industrial setting, is defined as 'fitness for purpose'. The 'purpose' of a product or system comes from the customer or customers, whilst the designers and manufacturers do their best to provide the 'fitness' (see Article 2.5). Putting these two aspects together is seen to be the route to the achievement of quality. This article spells out the consequences of this approach to quest for quality in engineering design.]

A good description of the design process is illustrated in Figure 1, which shows six core phases. These are basically in chronological order but are by no means watertight divisions; phases can merge into each other, and, as indicated by the smaller arrows, some phases can be repeated.

■ The identification of a need or want

This is the starting point of the design of a product or system. People's needs or wants may be established by a number of different means: for example, from a market survey performed to ascertain the likely demand, from direct contact with likely customers to discover the problems that they are trying to deal with, or, more riskily, just from a feeling that the intended product will sell. It is also possible to persuade people that a new product is desirable, even when no need for it exists, and in this way find a way to exploit a new material or process.

■ Specification

Once the purpose of the design has been identified, the limits and features of the final design must be defined in the design 'specification'. In addition to the

88

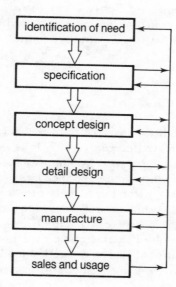

Figure 1 Design process model: core phases.

main performance characteristics which enable the design to fulfil its purpose, other factors need to be included in the specification, such as relevant legislation, patents, aesthetics, cost, producability, reliability, safety, manu-facturing route, etc. The specification is often the basis upon which a contract is signed. Once this specification has been prepared (and, where appropriate, the customer accepts it) the progress of the design can be measured against it. Sometimes it will prove necessary to change the specification as different data or new factors emerge or customer requirements change. Detailed specifications for different aspects or components are often drawn up and refined as the design progresses, but these must be within the overall specification.

■ Concept design

This phase can be started once the specification has been compiled. It is concerned initially with synthesis rather than analysis. Outline schemes are generated without worrying too much about details, whilst showing essential features and principles. The next stage is to decide which of the schemes should be developed into detailed schemes. To do this the outlines must be evaluated in some way. This can be done by simulation or by relying on experience or even by intuitive judgement. Each scheme must appear to be capable of conforming to the specification. It may be necessary to develop several schemes in parallel until one or more schemes are clearly better than the others. It may also be necessary to carry out research to find ways of achieving particular performance characteristics, such as safety or pollution control, etc.

■ Detail design

This phase is almost indistinguishable from the concept design phase, but it involves the addition of engineering detail. It is here that the engineering sciences can play their greatest part. Combinations and arrangements of any sub-assemblies are dealt with and the specifications of individual components in terms of shape, material, method of manufacture, type and accuracy of finish and so on are developed. While good detail design cannot usually rectify a poor concept, a well chosen concept can be ruined by poor detail design.

Much more is involved in this phase than just detailed drawing. It is likely to include calculations based on theoretical models of the proposed design as well as mock-ups to see if the design still fits the specification. The final check before the manufacturing of large quantities commences is the construction of a small number of prototypes, for evaluation by customers and sales staff, and for testing, possibly for long periods, in simulated or actual service conditions. The redesign of parts or even of the whole concept may ensue, though previous phases of the design process, especially the simulation and calculation phases, should have eliminated the possibility of this occurring.

■ Manufacturing

The intended manufacturing route has had to be borne in mind during all the previous phases, and may in fact have had a large effect on the choice between different outline schemes.

An important aspect of manufacturing is the quality assurance system used by the manufacturer. Quality assurance procedures are intended to ensure that all aspects of the manufacturing process, from the supply of parts on time to the quality of the working environment, are well maintained, and that the entire workforce is properly trained and continuously seeking to improve the quality of their work.

■ Sales and usage

All the previous stages have been aiming to ensure that the product is what the customer wants or needs, and that it is economically viable, maintainable, reliable and safe. Back-up in terms of servicing and maintenance information is very important, all of which had to be borne in mind during the preceding phases. Customers' reactions, how the product is actually used (and abused), performance and any unforeseen problems all need to be discovered. These data, as well as changes originating from the designers themselves, normally lead to improvements even in successful products.

Look again at Figure 1. The phases described are schematically represented. The division between the phases can become blurred. In fact a whole scheme of arrows connecting all phases would sometimes seem appropriate, although the route indicated by the heavy arrows in the diagram indicate the most effective way of proceeding. The feedback arrows represent the much less efficient process of iteration, and so represent a fall-back strategy rather than one to be deliberately pursued. Producing a quality product involves the successful completion of all the phases of the diagram, though in simple designs some stages can often be omitted. It is worth remembering that the success of the design process, like the strength of a chain, is no better than the quality of its weakest link.

2.3

Where are Designers Placed?

T263 course team

[This article, like the previous one, discusses the place of design in the context of a larger process, what the author here calls the 'product cycle'. It also represents some of the thought processes of designers as they work within this product cycle, and discusses how designers relate to others in the cycle. As the term 'product cycle' implies, the author is primarily concerned with product design.]

[. . .]

How does what designers do relate to what other people do and to other processes? Let's think of this in abstract terms first of all.

■ Design convergence

At the beginning of the design process all things seem possible to the designer. In one important sense anything is possible, because a designer can think anything. There are no prohibitions on what might be conceived. Moreover, there are only a few limitations (of skill and imagination) on what might be sketched, on what might form a beginning, a first explicit idea. This imponderably large range of conceptual possibilities is what makes design difficult. It is also the thing that makes it exciting.

However, at the end of the design process the designer has settled for one design: for one precise specification of a new object or circumstance. So, in this general sense, the activity of designing is framed between all possibilities and a single solution. This means that the designer is responsible for, and enmeshed in, a hugely convergent process. The beginning is all things; the end is one thing. [See Figure 1.]

92

At a more prosaic level, the designer may not have the freedom, or even the inclination, to explore more than a small number of possibilities. The territory would be mapped out beforehand.

[. . .]

Let's think about the beginnings and ends of the design process more realistically. Normally design would begin with a *brief*. The work of a professional designer is limited by some kind of definition. This brief may take the form of a highly detailed performance specification; a statement of goals to be achieved. It may even include outline drawings defining a mechanical configuration that is fixed. It may be a more general bureaucratic specification, as for a local authority's housing scheme. Or the brief may be some fairly casual prescription. For example, Eric Taylor, Design Director of Patterson Photographic, recalls his Managing Director saying to him, 'I was using these forceps last night, Eric. They kept slipping into the tray. I think we could do better than that.' Whatever the form of the brief, it circumscribes

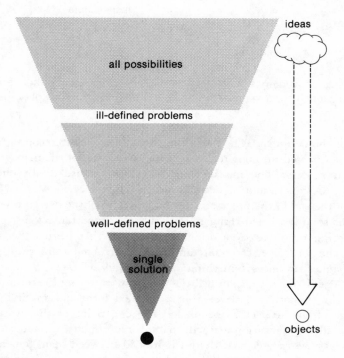

Figure 1   Design convergence in conceptual terms. A designer converts ill-defined problems into well-defined problems: a brief into a design.

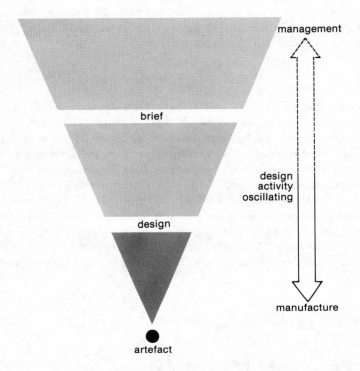

Figure 2   Design convergence in practical terms. A designer operates between
management and manufacture, with a good understanding of both.

the activity of the designer. While all things may seem possible conceptually,
the designer is directed into one territory. This is a matter of the management
of the design process. The management of design is usually beyond the
control of the designer, and sometimes beyond influence.

What of the end of the process? The end objective of the design process is
a design. This sounds a banal thing to say, but it is important to distinguish
between a design produced by a designer, the brief that begins the design
process, and the final artefact or circumstance constructed from the design.

The designer produces something which is only an *intermediate* step to
another end. The designer's responsibilities run from exploring the terms of the
brief, generating hypotheses, developing a detailed form and specifying the
precise nature of the artefact to be made. The end point of this work is a
technically detailed specification that will enable something to be manufactured.

In this way we can see that design is framed between management and
manufacture. The designer occupies the middle of this field of activities, neither
as a manager, nor as a manufacturer; although a good understanding of *both* is
necessary in order to design. [See Figure 2.]

This sounds rather formal, as if we are only to accept a definition of designers as *professional* experts: in part professional managers and in part professional supervisors of constructional processes.

But what if *you* design and make a bookshelf, or a chair for yourself? What if you knit a pullover to your own design, or plan and plant out your own garden? Clearly these activities entail design in some sense. They are akin to professional design. They require *some* of the abilities of professional designers. If you make a chair for yourself, or knit a pullover, or plan your garden, you have some understanding in your fingertips and in your head of what it means to be a designer.

However, if you are asked by a friend to make a pullover, or plan their garden, there is a significant shift in your role. You have a client. You have a brief. You would discuss the money to be spent, the general notions your friend has, their likes and dislikes, and then perhaps you would make one or two sketches. It becomes necessary to *externalize* the decisions that you might otherwise make for yourself without discussion. The process becomes more formal, and more dependent upon the communication of ideas through models and drawings. This is, in embryo, the work of the designer: discussing a brief, making a plan, offering tentative solutions.

You may go further: you may act on behalf of your friend, not only as the generator of ideas and plans, but you also may commission a skilled gardener, or knitter, to carry out these designs. Again you would step further into the role of a professional designer. You make a specification for a new object and you oversee its construction. You take up the designer's position in this intermediary zone between brief and manufacture.

These ideas can be represented diagrammatically. Look at Figure 3, the changing design process. Here we can see four kinds of design process. Firstly, an individual makes something in the way that you might make a chair, bookcase, pullover or garden. In this instance that individual is *simultaneously* the client, who provides the money, the user of the artefact, the designer, who tries out various ideas, and the maker, who actually constructs the thing itself.

Secondly, you may have a division of responsibilities between the individual who wants something and has money to pay for it, and the individual who sells her work. Our diagram subdivides into two roles, the designer–maker and the client–user.

Thirdly, Figure 3 represents a further division: a separation occurs between the job of having ideas and actually carrying out the construction. The designer becomes distinct from the maker.

Fourthly, we have the final separation. The person who provides the money and the commission to a designer is no longer the user of the designed object. Just think, for example, of how most mass-produced articles are designed.

[. . .]

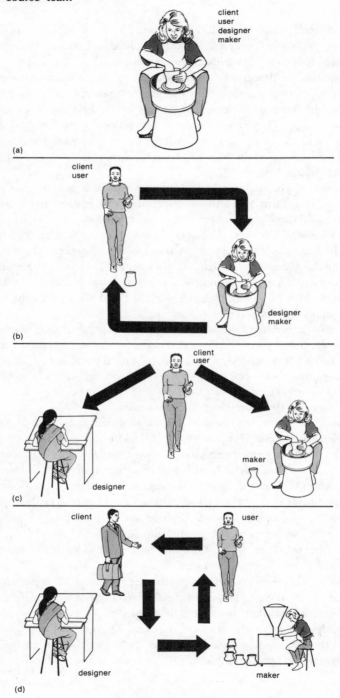

Figure 3   The changing design process.

■ The product cycle

[. . .] Our study of design has widened from the level of individual skills, to the connections designers make to other activities and other people. We have now arrived at a point where we can look at the design process in its social and technical context. By that I mean we are going to look at design as part of a bigger process and *not* just as the activity undertaken by designers. This larger process is generated by an unsatisfactory object or circumstance, a need which becomes pressing. A product is devised to fill this need, then manufactured, sold and used, until, in turn, it becomes worn out, useless, unsatisfactory or out of date. The process begins again.

Let's call this circular train of events the *product cycle*. Figure 4 is a simplification of this cycle. It may not apply to an industry you know well. You may like, after reading this [article], to modify the diagram, and description, to fit your special knowledge and experience. Figure 4 is intended as a framework; it simplifies things for us conveniently, but at the same time it disguises a few important issues.

Firstly, in its details the product cycle is *complicated*. Secondly, this process will *vary* from industry to industry, from company to company, from product to product. Thirdly, many *feedback loops* between adjacent groups and even between distant groups are not shown in Figure 4. Fourthly, certain important characteristics are *difficult to describe*, although I can point to them and talk around them. [There are] at least two instances of this resistance to formulation: in the internal creative processes that make some people more imaginative than others, and in the external qualities that make one design 'better' than another.

So what can be said? And what can be said that makes sense over a wide range of products?

We can start by *placing* people in their appropriate relationships to the design process. In the product cycle we can see certain identifiable tasks and roles, which connect with each other, so that the finishing point for one individual or group is the starting point for the next. Thus, in Figure 4:

A   In a group of **users** there arise certain specific wants and *needs*.

B   These needs are identified, and sorted out or stimulated, by **market researchers**, who formulate a *brief*.

C   This brief is subject to **Government** regulations and certain *standards* are enforced by law or by agreement.

D   The **client** will modify and implement the proposals of the brief on the basis of his *finance*.

E   The **designer** will react to the brief and provide a prescription for a new object: a *design*.

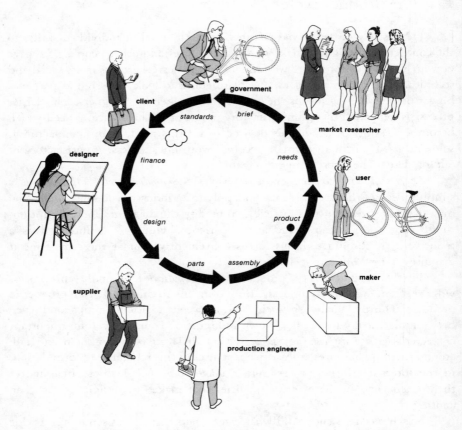

Figure 4   Product cycle. The main participants and their activities in the
promotion, design, manufacture and use of products.

F   Resources, materials and *parts* are organized from various **suppliers**.

G   The process of manufacture and *assembly* is organized by a **production engineer**.

H   Who sets up the conditions of work for the labour of the **makers**, who manufacture the *product*.

I   This product is then taken up from the marketplace by the **users**, who discover new preferences and new *needs*, and so we return to A.

Clearly this description is inadequate in many respects. It says very little about the process of distribution and retailing. It says little about needs that are synthetically created in the users. It says very little about the complex of advisers – accountants, materials experts, research workers, patent advisers and so on – that the designer calls upon. However, that does not matter greatly at this point.

What matters is that you understand three simple points:

(1) The whole process is a spiral. It feeds back to itself, it is continuous, but it moves forward in time (see Figure 5).

(2) Each person, or group of people, within the process sees that process differently.

(3) The starting point for one group is the finishing point for another.

Although there are many movements back and forth, discussions and debates, each group has its end-point and its own focus of attention. The focus of attention of each group is indicated roughly by the italicized word in Figure 4. Thus, needs become a brief, a brief is subject to standards, the finance makes the brief possible, the brief is drawn up as a task for the designer, a design is conceived as a production process, the engineers and makers convert conceptions into the facts of a product, the product suffers use and abuse in the outside world – and new demands become articulated.

If you think this model of the design process is rather like a hysterical game of pass the parcel with sudden reversals, then you would be right. But here the parcel is put together and *made tidier* as it is passed around. When the music stops, the users are left holding the parcel.

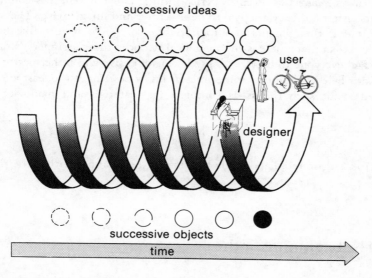

Figure 5 Evolution of design. Each revolution represents one turn of the product cycle, moving forward in time. Each successive idea leads to another form of the artefact, which leads to the next idea, and so on . . .

Figure 4, although crude, conveys the two important ideas that the flow of information is ultimately one way and that the designer is disconnected from the user. This disconnection can manifest itself in users' dissatisfaction with new products, as suggested in the following survey notes by the National Consumer Council: 'TV set; always need to have it repaired. I haven't done anything. They say if a TV is over three years old it is classed as an old TV.' 'Toaster would not do what it is supposed to do. We threw it in the dustbin.' (Source: National Consumer Council, Occasional Paper no. 1, *Faulty Goods*.)

On the other side of the divide, designers express dissatisfaction with users' judgements. For example, here two designers of consumer goods express dissatisfaction with the British public taste.

> Tony Worsfold, Nash Associates: 'The British public are undemanding, so manufacturers play safe. Look at the way they turned in new ceramic [cooker] hobs with old black-and-chrome trim, for instance. The result is we can't sell abroad.'
>
> Ben Fether, Fether and Partners: 'We've recently set up a system of researching our designs for consumer products in five European countries. The results we've so far received repeatedly underline the fact that the UK market, in terms of volume sales, is visually conservative, resists innovation, expects costs to be rock bottom and is indifferent to quality.' (*Design*, no. 380, August 1980, p. 43.)

These comments by designers are an illustration of the gap that can open up between professionals and the public for whom they cater. We could argue that it may be sensible in certain circumstances to be conservative. We could argue that an *unnecessary* (even hysterical) kind of inventiveness is common in designers. After all, they are paid to be inventive and imaginative. They would feel cheated if their clients and their public said, 'Do it just like before'.

More than that, there is an evangelical streak in most designers, who want to 'educate' everyone to better things. Is this offensive and rather patronizing? Or could it be that designers sometimes *do* know best, or at the least are better informed about such things as components, materials and manufacturing?

[. . .]

Let's think a little more about the location of the designer within the product cycle.

■   **The beginning of the cycle**

If the product cycle can be represented as a spiral, then it is hard to say at what point the process *begins*. Each person within the process *can* be the generator of a new thought, which may act as the stimulus for everyone else and lead to a totally new product. So where is the process likely to begin? It is unusual for the users to move forward from vague feelings of dissatisfaction. As a group they are not trained to focus on the source of their dissatisfaction nor to generate new possibilities. [. . .]

The initiative usually comes from the client or the designer. They may both operate within one company, or one seeks out the other. The client has the general notion and the money; the designer has the ideas and practical knowledge. Their initial work is against the background of users' dissatisfaction. As clever entrepreneurs or inventive designers they detect (or sometimes create) new demands. They become aware of shortcomings in a range of products, or in their own products. The power of their ideas and money puts them in a position to initiate changes.

The first moves, then, usually come from the client and designer. The source of new products tends to be ideas and enterprises from clients, designers or engineers, which manufacturers attempt to sell to users. This view of innovation is called the *technology-push*, or *design-push* model.

Alternatively, the source of new product can under some circumstances be the demands of the users, which became clear, articulate and pressing. Clients, designers and engineers then respond to these demands and manufacture articles to fill this need. This view of innovation is called the *market-pull* model.

A popular conception of how the process begins envisages a brain wave on the part of a designer: 'Eureka!' The brilliant inventor leaps from his bath. This is not *totally* wrong. A designer in a mood of relaxation may have a sudden inspiration, but two things need to be said by way of reservation.

Firstly, that moment of inspiration does not take place in a vacuum. It is preceded by an absorption of the facts, a manipulation of the main elements of the problem, a period of digestion and a turning over of these factors in the mind. The result is not so much a blinding flash of inspiration as a very clear perception of what the problem is and of how it could be solved. In retrospect it seems merely to be bringing to the surface something that was latent, and making the connection, however unlikely it seems.

'They are a very funny shape, you see. It's a question of should it be round. The first idea I had for the struts was to make them like a rhubarb stem, half circular, which is a strong section.' Here Ove Arup talks of Kingsgate Footbridge [. . .]

Secondly, that sudden inventive leap forward happens less often than you might think. It is an attractive and romantic idea. It emphasizes the creativity and brilliance of the designer; but much design is not like that at all. More often it is a matter of tinkering and adjustment to existing products. Designing becomes *incremental*. Instead of a sudden leap, we have step-by-step improvements. Instead of one designer's startling ideas, we have successive modifications from different people at different times.

This slower evolution of products is most likely to derive from the more severely practical members of the design team: from design engineers, fabricators, tool makers, production engineers, suppliers and specialist consultants. The focus of their attention is upon the product itself and the manufacturing process.

■ Thought and action

If we look at the simpler version of the product cycle shown in Figure 6, we can make some generalizations.

The participants in the cycle divide into those who are preoccupied with the *product* itself and those who are concerned with *information* to do with the product. A line drawn across the diagram separates researchers, consultants, economists, managers and entrepreneurs, who are concerned with the context and management of design, from engineers, constructors, the workforce, production engineers and so on, who are concerned with the physical properties of the product. The one faction is interested in *software*; the other faction is interested in hardware. Or, to put it in other words, one kind of person is interested in ideas, another kind is interested in practical details. If this makes them sound like people at war, well, very often they are in conflict.

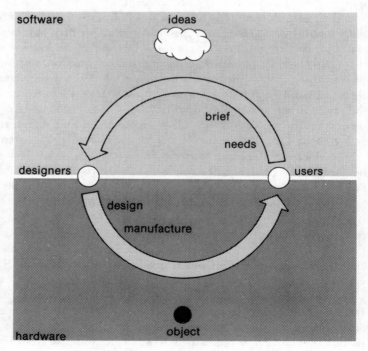

Figure 6   Ideas to objects. Product cycle shown as transition from software to hardware.

This is partly a matter of education and experience, and partly a matter of where they are *placed* in the design process. The one group is committed to *thought*: to reflection, to widening the possibilities and in general working with higher levels of uncertainty. The other group is committed to *action*: to construction, to narrowing the possibilities, and to a low degree of uncertainty. The pressures upon them are different. The one is preoccupied with demand, with what needs are to be met, the other with supply, with what material resources are available.

This diagram brings out another related point. There is a curious symmetry between designers and users. Designers translate one kind of information, a brief, into another kind of information, a design, which enables a product to be made. Users start from such products, from the hardware, and translate it into new demands and pressures. Designers make the translation from software towards hardware; users make the translation from hardware to software.

■ The moving designer

Let's examine in more detail that part of the product cycle that concerns the designer. As we have seen, designers are placed somewhere between a brief and a product, but they do not occupy a *fixed* position. Over a period of time they shift. At the beginning of the process they are closer to the client. The possibilities are more open. The designer may even take on the brief itself and revise the nature of the task. The problem may be reformulated in new terms.

Later in the process the designer's concerns are more to do with construction. The designer then is closer to the manufacturer. More than that, the nature of the problems has changed. Here we are deeper into the convergence towards one solution. The problems are focused and tightly defined. The designer's activity involves making rational choices between a few options. The processes of optimization can apply. Yet it is important to remember that the later processes with clearly defined problems and a few comparable options *depend* on the earlier processes of clarification.

In some circumstances the field of possibilities may have been restricted by the client. The brief may be quite specific, the possible options very reduced, and the designer is *not* required to challenge the assumptions of the brief. Much engineering design is straightforwardly prescriptive in this way. But what it means in general is that the client has taken over territory which in some other cases would be the province of the designer.

From [. . .] the description of the product cycle you can see that a designer needs to have a wide range of skills and abilities: on the one hand to be able to deal with out-of-focus imponderables, to challenge underlying assumptions and to generate alternatives, in short *to have ideas*, while on the other hand to be able to deal with tightly defined tasks, to make choices rationally and to prescribe technical specifications accurately, in short *to be practical* [see Figure 7].

Sometimes such individuals do exist, who can span the whole process of design from beginning to end. But this tends to be true only where the tasks of the designer are relatively simple, say designing furniture rather than designing motor cars. When the design task is technically complex then a *team* of designers becomes necessary. More than that, their specialities tend to be drawn to one side of the process or the other. The design team may split into two parts: those who operate best in the world of ideas, and those who operate best in the world of material objects. This is recognized by Richard Hamblin, who works with the private consultancy Ogle Designs.

'The managerial element means getting the right mix of people. We have people who are very creative but impractical – and people who are practical but without ideas – so we have to get them to work together.

The body stylist can get his ideas down on paper quickly, but that only solves part of the problem. It is only a beginning.' (*Richard Hamblin, 1981, in an interview with the author.*)

This polarity, having ideas or being practical, may be sometimes a false division, in that *some* people are both. Yet it relates not merely to the way

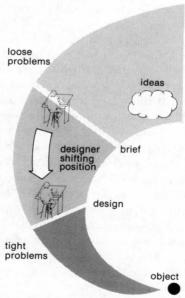

Figure 7 Design convergence within the product cycle. At the beginning of the design process the designer confronts loose problems and very many possibilities. At the end of the design process the designer achieves a tight definition of a formal, geometric and manufacturing problem, that is, a design.

design teams are organized, but to the way people think. [. . .] From our own observation we know that some people are able to operate best with open-ended problems, while others work best when faced with tight problems.

2.4

Design and the Economy

R. Roy and S. Potter et al.

[As the title of this article implies, the concern is with the aspects of design that are vital for the achievement of successful product development in a competitive environment. This article carries on the theme of the previous article, but stresses the need for companies to be economically successful. Although this view of design reflects industrial life, some aspects have found their way into technological activity in schools.]

■ The critical factors for success

Of the advanced industrial nations, Britain depends more than most on international trade to create prosperity and employment.

The competitive situation is difficult for Britain, as we have seen, and imports are penetrating further into domestic markets. Even those industries that are not concerned principally with exporting are confronted with competition from foreign firms and also from each other – and success lies in keeping up and getting in front of others in this competitive environment.

Research on design and innovation in British and foreign firms (e.g. NEDC, 1987) has shown the crucial importance of company culture and organization and of good management in achieving commercial success. Other studies (e.g. Roy, 1987; Walsh, Roy and Bruce, 1988) have shown that firms that win awards for good design and/or technical innovation perform better financially than other firms, but only if they are also generally well-managed and therefore good at other key aspects of their business such as strategic planning, marketing and manufacturing.

In this section we therefore examine the managerial and organizational factors that are vital for individual firms in achieving a successful product development, involving design and innovation.

'The performance and appearance of many could be substantially improved, which in turn would have the effect of increasing market appeal and reducing production costs. A poorly designed product may damage a company's reputation [. . .] particularly if the product is unreliable, unsafe, or difficult to maintain and repair. The benefits from improving product design are therefore considerable and, if they are to be realized, product design needs to be managed effectively.' ('*Guide to Managing Product Design*', *British Standard, British Standards Institution, 1989*.)

☐ Management of the design and innovation process

The development of a competitive product or process depends essentially on how well its market and technical specifications are defined and then translated into a commercially viable and economically manufacturable design.

This is the innovative design process. Thus innovation includes the technical, design, production, financial and marketing steps necessary for the first commercial sales of a new (or improved) manufactured product or the first commercial use of a new (or improved) manufacturing process. On the basis of previous failures and successes it is possible to identify eight essentially managerial issues that appear to be most important in the management of this design, development and commercial marketing sequence (Rothwell, 1977).

(1) *It is essential that management establishes good communication both with customers and with other external sources of ideas* (Rothwell, 1983). Few firms, especially small and medium sized firms, are entirely self-contained in their technological design expertise and the ability of these firms to 'plug in' to external sources of expertise and advice is crucial. Equally vital is the ability to communicate efficiently with the marketplace and to collaborate effectively with customers during design and development.

(2) *Innovation is not simply a matter of research, design and development. It is a corporate-wide task.* This highlights the importance of good internal co-operation and co-ordination between functionally separated departments. It implies that representatives of all research, development, manu-facturing and marketing functions should be involved from the very start of any project.

(3) Marketing and user need is the factor that has perhaps received the greatest emphasis on the various studies of innovation. *Successful innovators strive to match their product or process to the needs of the marketplace.* Too many would-be innovators have produced designs that were technically satisfying in themselves but which failed to meet the needs of potential customers, and this is a sure prescription for disaster. Determining user needs and satisfying them through good design is at the core of the successful innovation process.

(4) *Any development work must be efficient in the sense that technical bugs are*

eliminated before commercial launch. The literature is replete with examples of excellent design concepts that failed in the marketplace because of poor manufacturing and inadequate quality control procedures.

(5) Since innovation is a process of change, *the management of innovation is an extremely taxing undertaking that requires managers of high quality and ability.* Top management must be open-minded and progressive; indeed, unless top management has the will to innovate, there is little that other members of a firm can do to generate and expedite an effective innovation policy.

According to the classic work of Burns and Stalker (1961) and several recent studies (e.g. Pilditch, 1987), successful design and innovation tends to be associated with a participative management style and non-hierarchical, horizontal company structure, emphasizing consultation and a free flow of information rather than direction from above. Traditional methods of organization, which involve passing the emerging product from the marketing department to design to development to manufacture (and often back again) are no longer fast enough or suited to innovative product development. Instead multi-disciplinary team-work, or other methods of collaboration between marketing, design, manufacturing and other staff, is vital, to provide all the different skills required for innovative product development and fast enough to keep up with the competition.

(6) *In-house skills are vital.* Many firms will be compelled to utilize external expertise during the design and innovation process. However, this can never be a complete substitute for in-house skills. While a company might employ design consultants to assist in the creation of its new product, it nevertheless needs some in-house design expertise to enable it to capitalize on the external advice. Certainly many attempts at technology transfer have failed because companies did not acquire the necessary in-house skills (Rothwell, 1978).

Sometimes this was the result of the inability of the firm's managers to understand the commercial implications of the technical alternatives being offered by the consultants. At a more aggregated level, there exists a great deal of evidence to suggest that an important factor in British industrial decline is the fact that far fewer British managers have a background in engineering or design than those in competing countries such as West Germany (Lawrence, 1980).

[. . .]

(7) *After-sales service and user education are important.* With many new products and manufacturing processes, the responsibility of the manufacturer does not end with the commercial launch. When, for example, a complex piece of equipment is being used for the first time, especially by unsophisticated

customers, it is essential that training is given in the right uses and limitations of that equipment. Otherwise misuse and subsequent breakdown will, with some justification, be blamed on the equipment and not on the user.

Where possible, equipment should be designed to be 'user friendly'. It is also, of course essential that the manufacturer offers a speedy and efficient spares supply service. After-sales service and user servicing generally can be greatly facilitated if the equipment has designed-in serviceability.

(8) *The role of key individuals can be significant* (Rothwell, 1975). These key individuals are the so-called 'product champion' and 'business innovator'. The former enthusiastically supports the innovation, especially during its critical phases. The latter – the manager with overall control of the project – is also highly committed to the innovation and plays a key role in welding the different phases together into one continuous innovation process. This manager co-ordinates design activity during basic development, prototype production, and manufacturing. It is his duty to ensure that the product is designed with easy manufacture in mind and, most importantly, to make sure that at all times the focus of design activity is on the satisfaction of user needs.

☐ *A framework of support*

In order properly to capitalize on the activities and abilities of these key individuals, it is essential that the organizational framework enables them to operate effectively. The results of project SAPPHO, a study of success and failure in innovation in the chemical process and scientific instrument industries, highlighted the crucial importance of enthusiastic and committed product champions/business innovators and showed how company characteristics can influence their effectiveness (Rothwell *et al.*, 1984). For example, in the very large, hierarchical and sometimes bureaucratic firms in the chemical industry, the gifted and committed business innovator/product champion sometimes had his enthusiasm stifled by the weight of bureaucracy and his inability to get things moving. In this industry, commitment and enthusiasm had to go hand in hand with authority and power if the business innovator was to play his crucial role in success.

In contrast, in the scientific instruments industry, where the innovatory units tended to be small, it was the personal qualities of the business innovator – especially his enthusiasm and commitment to the innovation project – that were most important to success. In general, small units were seen to be the most conducive to the successful performance of 'in-house entrepreneurs'. This highlights the general problem of stimulating entrepreneurship in large organizations; large, rigidly administered and bureaucratic organizations tend to be intolerant of in-house entrepreneurs, whose innovatory endeavours are perceived as simply 'rocking the boat'. As a result, the would-be entrepreneurs either leave their firms or simply give up trying, a clear waste of valuable human capital.

[. . .]

Major Japanese corporations have demonstrated remarkable corporate flexibility during the past three decades. Commenting on this, Nicholas Valery (*The Economist*, 8 July 1983) wrote that 'this knack of being able to churn out new products with bewildering speed stems in part from the ease with which information is disseminated within Japanese companies. With their "bottom-up" organization, ideas move in Japanese firms from junior engineers to project managers and on to the boardroom directors to become corporate policy'. However it is achieved, it is clearly desirable that large companies create sufficient room within their organizations to accommodate entrepreneurship.

[. . .]

It is again worth emphasizing the role that design plays in industrial innovation in translating user needs into a set of functional specifications; in translating functional specifications into a functioning device; in linking product design to the manufacturing process; in making a product user friendly; in enhancing reliability; and in establishing high serviceability in use. In other words, the design and innovation processes in industry are so completely interlinked as to be inseparable.

Many of these comments on the management and organization of innovation are common sense. However, numerous studies and case histories have shown them to be vital for the success of a firm. Furthermore, when these management qualities are lacking the propensity for failure is substantial. The other factors that are most critical to success are all linked to the management of innovation, and they are only treated separately in the following pages in order to emphasize various points.

■ Identifying customer requirements

The obvious objective of any firm is to make things or to provide services that will sell profitably in their intended market, which means that it is vital for the requirements of potential customers to be understood and reflected in the design of the product.

The first and basic stage in either the improvement of an existing product or service, or in the creation of a new one, is to discover what the customer wants. Once this is understood it can be reflected in the market and product specifications used by the design team.

Providing 'what the customer wants' seems an obvious statement, but can be difficult to pin down. Good communications between end users and those responsible for the design of the products they will be using is absolutely critical. Design should not be a removed, exclusive activity. Some Japanese firms, for example, require their designers to sell their company's products as part of their training. However, anecdotal feedback can be unrepresentative and may be biased. So the following sections are intended to help in thinking through more systematically how customer requirements affect product design.

☐ Product characteristics

Marketers often talk about customer choice being determined by the 'four Ps' in the 'marketing mix' offered by companies competing in the market. These are **Product**; **Price**; **Place** and **Promotion**. Economists use a similar classification to describe the requirements affecting customer choice. These are more detailed, but consist of three groupings of **price factors**, **technical non-price** and **service non-price factors**. The following is a typical listing of these three groups.

☐ *Product characteristics influencing customer choice*

Price factors

- Price (list price; sales price; net price after trade-in allowance; etc.)
- Financial or leasing arrangements
- Lifecycle costs (whole life cost; allowing for running costs; servicing costs; breakdown costs; parts costs; depreciation; etc.).

Non-price 'technical' factors

- Performance in operation (e.g. speed and quality of operation)
- Reliability and durability
- Ease of use and maintenance
- Safety
- Appearance, materials and finish
- Flexibility and adaptability in use

- Packaging and presentation

etc.

Non-price 'service' factors

- Quality of after-sales servicing
- Delivery date
- User training facilities

etc.

'Price factors' include direct costs of purchase as well as the costs of ownership and use of the product. Some markets have a variety of prices depending on the type of customer and whether trade-ins are accepted. For some products the availability of leasing can significantly affect market demand. There is little doubt, for example, that leasing arrangements were a key factor in the rapid diffusion of the Xerox photocopier, since only relatively few offices could afford the high capital costs of the first models.

Leasing remains important in this market now for totally different reasons, because the technology is changing rapidly. For some products a lease and maintenance agreement may be the main market.

The 'non-price' factors show that the manufacturer is not just offering a piece of hardware but a package of technical and service characteristics. Customers will base their purchasing decisions on what blend is important to them and make trade-offs between various price and non-price factors. A new high speed combine harvester, for example, is a very poor buy, if, say, at harvest time it breaks down and spare parts are unavailable until November. Again, if

Table 1   How design affects customers' views of a product at different stages of purchase and use.

Phase	Product design factors
Before purchase:	Manufacturer's specification, advertised performance and appearance, test results, image of company's products, list price. ('brochure characteristics')
Purchase:	Overall design and quality, special features, materials, colour, finish, first impressions of performance, purchase price. ('showroom characteristics')
Initial use:	Actual performance, ease of use, safety, etc. ('performance characteristics')
Long-term use:	Reliability, ease of maintenance, durability, running cost, etc.

Adapted from: Walsh, Roy and Bruce, 1988

the servicing and cost of spares of a modestly-priced washing machine are inordinately high, it may be seen as a poor buy.

> 'The British medical equipment industry rarely looks at international markets. I have recently been commissioned for the first time by a German manufacturer, who has sent me to the United States, Japan – twice – and Europe to make sure that the product I am working on meets the particular market needs of those countries. In the twenty years that I have been working for British companies, I have never been asked to investigate overseas markets in this way.' (*Russell Manoy, Principal, Solari Manoy design consultants*)

It is the function of the manufacturer to identify an optimum set of characteristics that best match the requirements of a particular group of users. It is then the function of the designer to turn this set of user requirements into a working product or process that satisfies these requirements.

☐   *Intensity of desire for product characteristics*

As well as identifying a customer's desire for a particular product characteristic, it is necessary to estimate how intense that desire is. A customer may like an additional design feature, but is this liking intense enough for him or her to buy your product rather than a competitor's? For example, would a customer be prepared to pay an extra 5% for additional flexibility in the use of the product? So it is important to discover among the body of potential users just how much they are willing to pay to acquire each benefit that could be included in the product. Of course, the value of a benefit will vary with each individual user. For example, a new piece of production machinery might offer the benefit of 'high flexibility' that will be greatly valued by a small manufacturer who is involved in the batch production of a variety of products. However, another customer who produces large numbers of standardized products would not value flexibility highly at all. Both, though, are likely to place a high value on the benefit of improved reliability.

One way to approach this situation is to identify a collection of user requirements that would satisfy a 'broad set' of users. Estimates can be made of the intensity of each requirement. How this is done can vary from sophisticated market research models through to an informal chat with sales staff. This would produce a set of 'high intensity' requirements that must be met in the design and a set of 'low intensity' requirements that might be met if they can be incorporated into the design without taking up too much time or cost. These can then be incorporated into the design specification as a list of design priorities.

A variation on this method occurs if you have clear groups of customers with a different set of user requirements intensities. Under such circumstances you should seriously consider whether a single product designed to satisfy a 'broad' set of users might simply fall between stools and not satisfy any of them. In such a case you could: (a) split the product into a range, with each product in the range designed for the requirements of the customers in that sector, or (b) if it is viable, provide a degree of customization for your product.

[. . .]

☐   *Stability of user requirements*

A further factor is that such intensities may change over time. It was not very long ago that 'environmental friendliness' would have hardly featured as a priority in consumers' buying preferences. Today it is the selling point of many goods. A firm that fails to notice this sort of preference shift can find itself outclassed by competitors who design for new and changed situations. A revised set of priorities is required; the design process must be dynamic and continuous. Re-design and re-innovation are essential to keep up with changing market preferences.

One point worth stressing is that market preferences may change in the period between the original design specification and launch. Speed of product development is becoming increasingly important and user requirements now change rapidly, often in response to new product developments. The designer must continually update the specification to meet any changes in market demand.

Even with relatively mature products, the intensity of user requirements can change significantly. For example, the rise in oil prices in the early 1970s suddenly resulted in the fuel economy of cars becoming a far greater consumer priority than it had previously been. Firms such as Chrysler US that did not adapt to this change in intensity as quickly as their competitors lost out commercially.

In Britain in the 1970s and 80s, the rise of the company car sector has resulted in a distinct split in consumer preferences. The subsidized company motorist is little interested in fuel economy or maintenance costs but is very concerned about specification and performance. A private motorist, who has to pay for his own fuel and maintenance bills, has a very different view! Ford quickly recognized the important intensity differences of the company car sector and developed designs specifically aimed at the company motorist. They now dominate company car sales.

> 'Time is of the essence. Today, companies make history, or are consigned to it, quickly. In many industries, six months can be packed with moves and countermoves. Products are born, sold, and phased out. Information moves very quickly. Customers will not wait; indeed, they will pay a premium for responsiveness . . . Companies that shorten product development time often prompt associated improvements in product quality and lower development costs. Their personnel

become better at knowing what *"better"* is.' (*Kim B. Clark, Professor of Business Administration, Harvard Business School*)

With the rise of 'green' issues, is this set to change again? Companies may wish to present an 'environmentally friendly' image through the cars that their managers and representatives use. Unleaded fuel may only be the first step. . . .

Another factor with mature products is that, as primary requirements are met in product design, secondary requirements take on a new significance. Today all modern cars are expected to have the design features that make them a reliable form of transport. The focus of competition has therefore shifted to matters such as better seat design, ergonomic improvements in instrument layout, soundproofing and the quality of sound obtained from the in-car audio equipment.

This example again emphasizes that designers are aiming at a moving target and that firms must remain 'plugged in' to their customers, continuously monitoring and even anticipating changing user requirements. There is no room for complacency in the competitive design process.

[. . .]

■ The role of the user in product design

One way in which the assessment of user requirements can be helped is by close contacts between the manufacturer and the users (or potential users) of a product. Obviously there will be customer feedback via sales staff, or simply in terms of what does and does not sell. But such information only influences product design after launch, and is often selective and vague. More direct user links are quicker and can focus down on crucial design issues. In today's fast moving competitive environment, this can be the difference between success and failure. Particular benefits of close user involvement include:

- complementing your own R, D&D abilities through plugging into the technical strengths of your customers
- involving the user is an aid to establishing the optimum performance/price combination, which in turn establishes the optimum design specification
- and involving users can result in a flow of user-initiated improvements to the original design.

The first situation certainly applies to smaller firms who may have only a very limited scientific, technological and design expertise. However, this situation can apply to large companies too.

Turning to user involvement in basic design and innovation, a study of 25 commercially successful textile machinery innovations showed that 84% of the companies concerned collaborated with people outside the firm itself and that 66% of these collaborations were with potential customers. Two-thirds of the collaboration took place at the R&D stage, most of the rest being at the production stage. Only a tiny proportion (under 5%) left collaboration with users until the marketing and sales stage.

A classical study of the involvement of users in the design and innovation process was von Hippel's researches in the United States in the late 1970s. He found that out of a total of 160 innovations in the scientific instruments and electronic manufacturing equipment industries, 117 were invented by users (von Hippel, 1978). Shaw (1983) found a similar pattern in the medical instruments industry: in a sample of 33 innovations, 25 were transferred from the users to manufacturer via a process of continuous interaction. Significantly 22 of these were commercially successful and only two failed (due to poor engineering, not because the design concept was wrong). The other innovation was too early in its life to be judged.

Users can also play a crucial role in 're-innovation'; i.e. in the improvement of product performance following its first commercial launch. Moreover, in some instances, innovations can be designed such that they are amenable to user re-innovation (Rothwell, 1986).

However, it is important to have a close liaison with the right type of user. Consider, for example:

- Does the user have a track record for purchasing up-to-date equipment and utilizing it appropriately. Liaising with a behind-the-times customer won't do much good!
- 'Tough' demanding customers are known to stimulate superior designs that represent good value for money and so win high sales. (Gardiner and Rothwell, 1985, explore this point.)
- Is the user typical of the market? Or is there a need to choose several users to cover different market segments?

[. . .]

Exports represent an important market segmentation. Holt (1987) provides a telling quote from the founder of Sony, Masaru Ibuka, who stresses how important it is to consider national differences in design requirements:

'Japan is not a closed market, but you Westerners don't try hard enough. You make no effort to sell to us. You don't learn our language. Your car makers don't even place the steering wheel on the right side when you try to export cars to us. In contrast, when we decide to enter a new market with our products, we make a whole-hearted commitment. Every country has its own peculiar characteristics, and we adapt our products accordingly. We study and examine the market, and we struggle hard.'

On the basis of this evidence the often quoted slogan for would-be innovators 'find a need and fill it' should have added to it: 'and actively involve innovative customers in the process'. Certainly in many industries the innovative user has a crucial role to play in the design specification of a product and in the continual improvement of that design. Supplier companies ignore this at their peril.

■ Using external innovative design ideas

Few firms are completely self-contained in research, design and development experience. There are a lot of external sources of potentially useful ideas and advice.

[. . .]

One disturbing feature of British industry is the 'not invented here' syndrome, whereby ideas, designs and innovations originating from outside the company are dismissed or treated with suspicion. A study of British companies and 'world leading' foreign firms noted a distinctive difference in attitudes to external design ideas. The British companies relied far more than the foreign firms on their own design ideas rather than ideas originating from suppliers or competitors (Design Innovation Group, 1990). [. . .]

The use of external consultants contains similar lessons. Very often a considerable amount of time can be saved by exploring an innovative idea with a specialist. For example, an electronics company wanted to explore the idea of using optical fibres rather than conventional wiring for a marine instrument. They had no expertise in this area and engaged a specialist consultant who quickly showed that, although for many applications optical fibres now have an advantage, the hostile operating environment of the open seas is not one of them. The conclusion was a negative one, but in a competitive market, it was an avenue they felt they had to explore. Some areas of specialist knowledge are such that it is not viable to employ somebody full time in-house, because they would only occasionally be used. This is true for many specialist engineering design skills, but can also be true of industrial and graphic design for smaller companies.

[. . .]

■ Designing for appearance, ergonomics and manufacture

So far we have tended to underplay the role of design in improving competitiveness in three crucial areas:

- giving new or updated products an attractive visual form, style and image
- ensuring that new or updated products are easy and safe for all users
- ensuring that new or updated products are economical to manufacture.

These aspects of design are not something to be added on at the end of the product development process. They should be considered right from the start and be included as a part of the brief and the product specification. It means involving people with the relevant expertise – e.g. industrial designers, ergonomists, and production engineers – throughout product development.

It is also worth noting that improved technologies and materials can allow the design of new forms and styles. For example, electronic equipment can be designed to be much smaller, lighter and in a greater variety of forms than in the past due to new components and materials.

□ Design for appearance and ergonomics

Other things being equal, customers will prefer a product that appeals to them visually and which is easy and safe to use. Aesthetic appeal is obviously one of the most important selling factors for products such as clothing, toys, ceramics, furniture and textiles. But styling, colour and finish are also more important than is usually appreciated in selling goods ranging from scientific instruments and computers to tools and vehicles of all types. For example, Moody (1980) has shown that successful manufacturers of scientific and medical equipment appreciate that one of the key factors in customer choice is elegant visual design.

Research on the costs and benefits of investment in design [. . .] has provided many examples of British products whose commercial performance has greatly improved following redesign involving professional design consultants to improve appearance and/or ergonomics. One example is the redesign of the desk and cladding of a sound mixing console used in recording studios. After the replacement of the old-fashioned wood and steel structure with a modern plastics moulding, sales of the mixer more than doubled and its profit margin and share of the world market also increased.

Another example is the redesign of a British crash helmet to compete with the 'flair and style' of imported Italian designs. Sales of the new design were nearly three times those of the old and the company also benefited generally from having a more modern image in the market.

☐ Design for manufacture

However well designed a product is in terms of performance, appearance, ergonomics, etc., it cannot compete if it is expensive or difficult to manufacture. One very important way of enabling economic manufacture is to ensure that production facilities and skills, choice of materials and methods of assembly and so on are fully considered in the design of the product and not passed to the production department to sort out at a late stage. The Ford Motor Company estimates that design decisions are ten times as effective as production planning decisions and one hundred times as effective as production changes in reducing costs and improving quality.

Design for low cost, high quality manufacture should be part of a total, multi-disciplinary approach in which all the user and other design requirements are taken into account.

[. . .]

[An] example concerns the design of a railway coach for the world market. In the 1980s British Rail Engineering Ltd (BREL) created a design using modular construction and a standard kit of parts that could be built on a simple system of moveable jigs and tools. The design of the product, and the associated flexible manufacturing system, enables coaches of different lengths, widths, heights, interior layouts, equipment options, etc. to be economically built on the same production line. BREL's International Coach, and the technology for making it, has been exported to several countries round the world (Roy, Walker and Cross, 1987).

☐ Design families

Designing for manufacture frequently leads, as in the case of the International Coach mentioned above, to a modular design and method of construction that permits a 'design family' of products to be made.

The design family can be up-rated, re-rated, or down-rated to suit a particular user's requirements. The importance of this design philosophy holds not only for large firms but for small firms and for high or low technology products.

One area in which Britain has been particularly adept at producing highly competitive and successful design families is in high powered aeroengines. The development of the large fan, high by-pass ratio engine led to the emergence of three basic design families which between them now account for over 50% in value of the Western commercial aeroengine market. These three design families are the Rolls-Royce RB211 family, the General Electric CF6 family and the Pratt and Whitney JT9D family. From the late 1960s to the present day, each of the manufacturers has developed a range of versions of its basic engine to meet a variety of user needs. Looking at the RB211 family, it is notable that the engines have just seven major sub-assemblies, thus greatly facilitating modular production and construction as well as greatly improving ease of servicing and maintenance for the user. Modular construction has also provided the design and production flexibilities that facilitate the creation of a growing family of up-rated, de-rated and re-rated engines (Rothwell and Gardiner, 1989).

■ Radical and incremental innovation and design

A marked omission in some studies of industrial innovation and design is the failure to recognize that major innovations will usually be subjected to continuous modifications and improvements. Over many decades, the sum of quite minor modifications can often have as great an overall impact as did the original breakthrough. A good example of the role of re-innovation is the television set. Following the initial technological breakthrough, the television was characterized by a small screen, poor quality, frequently unstable pictures and was prone to electrical breakdowns. Through incremental design improvements and the use of better components, modern television sets now have large screens, fewer parts, better quality pictures and greatly improved reliability. Furthermore, all of this is available at a much reduced real price.

While design is essential to the translation of even a major technological breakthrough to commercial use, it is in the study of these subsequent incremental innovations that the influence of good design on performance, and hence on competitiveness, can most easily be detected [. . .]

Another example, the success of the Ford Cortina in the 1960s and 70s, was not so much due to its original, somewhat conventional design, but because it had been designed to evolve and adapt continually to different customer requirements. It was thus able to accommodate easily changes in market preferences and incrementally introduce technical improvements. British Leyland's 1100/1300 series cars, introduced at the same time as the Cortina, were less commercially successful over their lifetime despite being more innovative in design. They featured, for example, a transverse engine, front wheel drive and 'hydrolastic' suspension. The more sophisticated 1100/1300 series was less capable of being evolved and varied to satisfy different markets and changing customer requirements (Rothwell and Gardiner, 1984).

This does not mean that incrementalism is always preferable to radical change. A radical breakthrough, as in the case of the Harrier jump jet, can be the key to opening up totally new markets. But keeping those markets needs continual improvement. Take the Sony Walkman personal stereo player. This radical concept was only possible through the earlier innovation of cassette tapes and miniaturized electronics to which was added specially developed light-weight stereo headphones. But although Sony created the Walkman market, it was not long before almost every other Japanese consumer electronics company had produced its own design of personal stereo. To keep up with this competition, Sony has had to continuously improve its own designs and develop a range of different models for different segments of the market, including unusual types like the water resistant, yellow, Sports Walkman and even a solar powered model (Roy, Walker and Cross, 1987).

The lesson seems to be that while good new designs can be based on either traditional or radical technologies, long-term commercial success depends on their continuous improvement. [. . .]

■ References

Burns, R. and Stalker, G.M. (1961) *The Management of Innovation*. London: Tavistock Publications.

Design Innovation Group (1990) *Winning by Design*. Oxford: Basil Blackwell.

Gardiner, P. and Rothwell, R. (1985) Tough customers: good designs. *Design Studies*, 6 (1), January, 7–17.

von Hippel, E. (1978) The dominant role of users in the scientific instrument innovation process. *Research Policy*, 5, 212–36.

Holt, K. (1987) *Innovation: A Challenge to the Engineer*. Amsterdam and New York: Elsevier Science Publishing.

Lawrence, P. (1980) *Managers and Management in West Germany*. London: Croom Helm.

Moody, S. (1980) The role of industrial design in technological innovation. *Design Studies*, 1 (6).

NEDC (1987) *Design for a Corporate Culture: A Report Prepared for the National Economic Development Council by James Fairhead*. London: National Economic Development Office.

Pilditch, J. (1987) *Winning Ways*. London: Harper and Row.

Rothwell, R. (1975) Intracorporate entrepreviews. *Management Decision*, 13 (3).

Rothwell, R. (1977) Characteristics of successful innovators and technically progressive firms. *R&D Management*, 7 (3).

Rothwell, R. (1978) Some problems of technology transfer into industry: examples from the textile machinery sector. *IEE Transactions on Engineering Management*, February.

Rothwell, R. (1983) *Information and Successful Innovation*. Report No 5782 prepared for the British Library, Research and Development Department, London, February.

Rothwell, R. (1986) Innovation and re-innovation: a role for the user. *Journal of Marketing Management*, 2.

Rothwell, R. *et al.* (1984) SAPPHO Phase II. *Research Policy*, 3 (3), 258–91.

Rothwell, R. and Gardiner, J.P. (1984) The role of design in competitiveness. In Langdon, R. (ed.) *Design Policy Vol 2, Design and Industry*. London: Design Council.

Rothwell, R. and Gardiner, J.P. (1989) The strategic management of re-innovation. *R&D Management*, 19 (2).

Roy, R. (1987) Design for business success. *Engineering*, January, 16–17.

Roy, R., Walker, D. and Cross, N. (1987) *Design for the Market*. Watford: Engineering Industry Training Board Publications.

Shaw, B. (1983) The role of the user in the generation of innovations in the UK medical equipment industry. *Workshop on Innovation Process in Health Care: Historical and Contemporary Perspectives*. University of York, 23–25 March.

Walsh, V.M., Roy, R. and Bruce, M. (1988) Competitive by design. *Journal of Marketing Management*, 4 (2), Winter, 201–16.

2.5

Quality and Quality Assurance

J. Sparkes

■ Introduction

The quest for quality has recently risen to near the top of the agenda not only in all fields of industry, but in such fields as education and health care too. As Lord Sieff, chairman of Marks and Spencer put it, 'Quality is the fastest growing commodity in the world today'. So the twin problems of what 'quality' is, and how to achieve it consistently are the subject of this article.

Although politicians, planners and other people in the public eye are invariably as in favour of quality as they are of motherhood, they rarely take the time to explain what they mean by it. Certainly everyone has a vague idea of what it means, but it seems to be a rather elusive concept. To quote Persig in *Zen and the Art of Motorcycle Maintenance*:

> 'But when you try to say what quality is, apart from things that have it, all goes "poof". There's nothing to talk about. But if you can't say what quality is, how do you know what it is, or how do you know that it even exists?'

So the next section is concerned with clarifying what quality means, at least in the context of technology.

The third section is concerned with the problem of how to achieve quality consistently in a technological enterprise. As explained later, there are two basic strategies for doing this: one is usually called 'quality control' and the other is called 'quality assurance'. This article is mainly concerned with 'quality assurance'.

■ The meaning of quality

First, I think it is worth noting that the term 'quality' really means 'good quality'. Obviously, it is possible for products to be of 'poor quality' or of 'good quality'. The important quest is evidently for 'good' quality. But this has become shortened to the 'quest for quality', it being taken for granted that good quality is the aim. But what is meant by quality, or good quality?

Quality in products used to refer only to the style of the design, to the use of expensive materials and to the excellence of the workmanship, so that it was possible for an expert designer to look at a product and decide whether it was of high quality or not. Indeed, where antiques are concerned, this is still essentially what quality means. Nowadays, however, where new products and services are concerned, it has a broader and deeper meaning, which includes not only the excellence of the product, but also a consideration of the needs of the customer and the effectiveness of the process by which the product or service has been created. In other words, 'quality' now refers as much to the process of producing and designing a product or providing a service as to the nature of the final outcome. A key aspect of this new interpretation of the meaning of quality is its emphasis on the needs and/or wishes of the customer, whether 'the customer' is the next person along a production line who receives your 'product', or the person who eventually buys or receives what is on offer. Quality is therefore not primarily a matter of being approved by experts, it is also, and very importantly, a matter of providing what the customer asks for or is perceived to need.

This idea of the importance of the customer was expressed quite clearly in an advertisement for Nissan cars, which declared that their cars were:

> '. . . being built in Britain using a unique approach to precision manufacture where every worker is personally responsible for quality, with the next worker down the line acting as his customer.'

Indeed, as we shall see, this brief sentence encapsulates some of the main ideas of both the meaning of quality and the process of quality assurance.

This broader idea of quality is thrown strongly into relief by considering the needs of Third World countries. They often can't use some of the 'high-tech' products of industrialized countries because they lack the necessary complex infrastructure of roads and service stations, trained maintenance staff, etc. The supply, some years ago, of first-rate tractors to Kenya and other African countries, in order to help them increase their food production, was soon seen as a mistake, because they did not also have the workshops and the expertise to repair the machines when things went wrong. This doesn't mean that they should be palmed off with the rejects of the more affluent societies; they need quality products, too, but quality of a quite different kind, one that is matched to their particular environments. For example, it was realized that, while satisfying a real need, products should in many cases be 'low-tech' and easily maintained. At the same time, they must also be reliable, durable, safe and cheap. If designs for such a market achieve these characteristics they are entitled to be called quality products, even if they would not be so called in a different culture and environment.

The same principle applies everywhere, not just to Third World countries. The Japanese have penetrated British markets, despite a reputation for cheapness before the war and despite the horrors of the war itself, because they provided what people wanted, both at a price they could afford and in terms of many desirable performance factors. They were 'quality' products in the present meaning of the word, though perhaps not in the sense that traditionally singles out the Rolls-Royces of this world.

Meeting customers' needs is not however as straightforward as simply asking what they want, if only because customers have different opinions on the matter. But, in any case, people rarely know so precisely what they want that their views can be regarded as a complete specification. A better approach is to find out the kind of problems or activities that likely customers are encountering, and then to use one's expertise to provide a good solution. And in order to cope with differing opinions, an obvious strategy is to provide a variety of solutions, such as the range of cars available from a manufacturer — each of which in its own way is of high quality, but meets the needs or wants of a different set of customers. The main implication of this is that to achieve quality products, suppliers and customers should get together in a collaborative frame of mind to define quality in each context. The producer's own expert judgement is no longer enough.

This approach to defining quality can be briefly summarized as

Quality = fitness for purpose

The customer indicates the 'purpose' and the provider ensures the 'fitness'.

More formally, quality is defined by the British Standards Institute (BS4778) as: 'The totality of features and characteristics of a product or service that bear on its ability to satisfy stated aims or implied needs'.

But even though the general definition of quality is fairly straightforward; applying it to particular kinds of products can still be quite difficult, and implementing it is still full of problems. With manufactured products it may not be too difficult to assess whether some specification criteria have been implemented, because the criteria can often be expressed in terms of clearly measurable properties — such as size or weight or power output or colour, etc. But with the more subjective criteria, such as those relating to the quality of services like health care or education, the performance may not be measurable in the same way.

More importantly, however, experience shows that it is not in any case possible to test for quality of a product simply by measurements of its performance when completed. Even if all the failures have been rejected at the end of a production line — so that those accepted have passed all the tests — there is no guarantee that the next specimen will not fail. In other words, there is more to achieving quality than applying a series of tests to a finished product. It is also necessary to design quality into the product through an understanding of failure mechanisms and of how to achieve the characteristics that the customers regard as important. Trial-and-error methods, or even 'trial-and-success' (i.e. making something and finding it to be satisfactory) are only last resorts when there are no relevant principles – of science, technology, economics, etc. – to learn, understand and implement. Even if the feedback from the customers is favourable during a trial, this is no guarantee of future success with different customers and in different circumstances if the reasons for success are not understood. Equally, just knowing that the performance was *not* satisfactory, however detailed the information, does not itself indicate how to put it right. The causes of failure must be understood if at all possible. Evidently, then, although feedback from customers is important because errors can always occur, it is not in general enough to ensure quality. Quality must be 'built in' as well as 'designed in'. To achieve this kind of quality usually needs the full cooperation of the workforce, as explained in the section on quality assurance below.

Evidently then the primary need, wherever possible, is to *understand* how specified goals can be achieved so that the appropriate processes can be put in place. Only then, when the goals have been established and specified, is it possible to design quality into both the 'product' and the 'processes'. But complete understanding depends on experience as well as on grasping the relevant underlying principles, so it is always necessary to learn from feedback and from trial-and-error or trial-and-success. Building in the findings of experience during production is sometimes called 'feedforward', or simply the application of 'know-how'.

The importance of 'understanding' in technology, even if it is not so important in other fields, can be seen by considering what happens in the event of a disaster. If it is subsequently found that well-established principles have been ignored in the design or production of the product or system, the designers and producers can rightly be accused of negligence. The argument that such a disaster had never happened before is no excuse. The trouble with relying on feedback to ensure success, important though it is, is that it occurs *after* a failure rather than before it, so it can never prevent the first disaster. After an aircraft accident one doesn't just try a different design and hope for the best, experts look for the cause and try to understand how it occurred before flights are resumed. Even when well-established understanding is incomplete it is essential to deploy what there is, rather than rely solely on feedback or trial-and-error.

It is tempting, like Persig, to say that because we haven't got a wholly watertight definition of quality it is pointless to try to say anything positive about it. Equally, in fields such as education and health care, because we haven't complete theories of how to achieve success, it is tempting to say we have to rely only on the experience, and keep hoping for the best. But a moment's thought reveals that even incomplete understanding, so long as it is well-established, provides a better platform from which to make progress than relying on trial-and-error alone.

■ Quality control

Whatever definition of quality one adopts, there is a need in any enterprise to ensure that the aimed-for quality is being achieved. This is true whether the aim is to make cheap products or to use the best materials and employ the best workers in the business. One strategy for doing this has come to be called 'quality control'.

The British Standard definition (BS4778) of 'quality control' is: 'The operational techniques and activities that are used to fulfil requirements of quality'. In practice, quality control has come to mean the method of monitoring and controlling production through the use of 'inspectors'. The inspectors check the quality of items at various stages in the production process, as well as the quality of the final product coming off the production line. They reject, or return for reworking, any faulty items they find and feedback information to the production engineers about processes that might be improved. They cannot usually inspect every item that is made, so they have to test small samples and decide, by statistical analysis from these batch tests, whether the production process is properly under control, or whether too much variation in output is occurring. All this seems obvious enough and can be a successful way of maintaining quality if managed very skilfully, but problems arise, as follows:

- It tends to cause alienation of the workforce, and to create a 'them-and-us' attitude within the company.
- Inspectors can make mistakes just like the rest of us.
- Because the workforce is not responsible for quality, the people involved tend to lose interest in the product, become bored, and become more concerned about 'getting their work past the inspectors' than about the quality of their work.

Quality control like this has a long history. Inspectors began to make their appearance on the industrial scene around the time of the First World War. Before that, inspection was performed by the workers themselves and by their foremen. In 1923, an American engineer, F.W. Taylor, published a book on the so-called 'scientific approach to management' which led to a massive search for efficiency and control. He showed that by using inspectors to measure the output of really capable workers he could use these as standards against which the output of others could be compared. The difference between the two would represent the scope for increasing the efficiency of the less efficient workers and thereby also for raising their earnings. In a now famous experiment at the Bethlehem Steel Works in the US he was able not only to decrease the number of men needed to load wagons from 500 to 140, but also to increase their daily earnings by 60%. Piece-work payments (i.e. payments related to the quantity of work done rather than to the time worked) were used to motivate the workers to meet the standards that had been set. Taylor's original ideas were mainly concerned with the *quantity* of work achieved, but inspectors were later expected to monitor *quality* as well. The whole system became known as 'Taylorism'.

It soon became apparent, however, that although it was possible by sympathetic management to apply Taylorism successfully, a too rigid adoption of the system led to the problems listed earlier, with consequent absenteeism and a higher turnover of labour. So while many companies tried, and are still trying, with reasonable success, to make Taylorism work, others were attempting to develop better methods. These methods are now commonly referred to as 'quality assurance'.

■ Quality assurance

Quality assurance is defined as follows in BS4778. This definition is now adopted internationally. Quality assurance is: 'All those planned and systematic actions necessary to provide adequate confidence that a product or service will satisfy given requirements for quality'. To which a note is added: 'Unless given requirements fully reflect the needs of the user, quality assurance will not be complete.'

It is not surprising, with so non-specific a definition, that there are now a number of different interpretations of it, each of which claims to be called 'quality assurance'. The main contenders are:

(1) The British Standard BS5750: 'Quality Systems', which is identical to the international standard ISO 9000, and

(2) Total Quality Management (TQM).

Very briefly, BS5750 specifies a number of actions which a company seeking approval must take in order to achieve minimum acceptable standards, while TQM emphasizes the importance of training, of giving responsibility for quality to the people doing the work and of the importance of the continual search for improvement. The difference has been succinctly summarized as:

TQM is a religion; BS5750 is the law

In general, out of a maximum score of 10 for perfect quality assurance, the implementation of BS5750 will give a company a score of about 3/10. But by the application of TQM in addition, the best companies achieve scores of perhaps 8/10. So BS5750 starts one off, and establishes a minimum standard, but it doesn't get one very far.

The essence of quality assurance, as interpreted in TQM, is the idea that quality can only be achieved consistently and continuously if those doing the work are interested in it and are given the opportunity to suggest improvements, bring them about and obtain recognition for their ideas. It therefore regards training as a crucial element in the overall system: training in both the latest ways of doing their job, as well as in the methods of quality assurance and the skills of monitoring their own work. In TQM, inspectors become advisers and trainers rather than a regulatory arm of management. In other words, quality control becomes self-imposed; it is not something imposed from above, or looked after by a separate group of employees. It becomes the concern and responsibility of everyone involved in the enterprise, even though it has to be monitored too. Management has the responsibility of finding ways of ensuring that this attitude is universally accepted in the company.

On the other hand, the essence of quality assurance as interpreted by BS5750 is the putting in place in a company of a number of procedures, which together amount to a 'quality system' defined by the BSI as: 'The organizational structure, responsibilities, procedures and resources for implementing quality management' (BS4778).

BS5750 lists 20 procedures which are an essential minimum for any company wishing to comply with its requirements. These include an extensive documentation of production methods, as well as review procedures which make it possible for external auditors to satisfy themselves that an effective quality system is in place. The proponents of TQM argue that BS5750 is more in line with the definition of 'quality control' and is too bureaucratic. Proponents of BS5750 argue that TQM is not a 'system' and relies too much on good will. Both arguments have some validity. However evidence seems to suggest that, in general, more is required of a company than the implementation of BS5750 if it is to 'provide adequate confidence that a product or service will satisfy given requirements of quality'. And it is probably because of its emphasis on more than a prescription of well-defined processes that the term 'quality assurance' has come to be more associated with TQM than with BS5750. Evidently the difference between TQM and BS5750 is not just a difference of technique, it also reflects a difference of attitude and culture within an organization. Nevertheless, despite this difference, companies are finding it possible to put the requirements of BS5750 in place, and then, through TQM, to develop much better quality products and processes.

How then can companies bring about TQM if it is felt that a change to this kind of quality assurance is needed?

The first thing to accept is that the introduction of TQM takes time. Since it often requires a change of attitude in the company, particularly a belief that management and the workforce have the same aims, and that both rely on the other if success is to be achieved, the immediate introduction of TQM is impossible. Indeed, it is essential that the whole philosophy of quality assurance is accepted at board level, and that it is not just used as a way of increasing productivity and profits – even though these usually result. The following statement issued by the President of Philips NV, the multinational electrical and electronics company based at Eindhoven, is an example of how such changes can be initiated.

'The Board of Management has decided to give vigorous direction to a Company-wide approach to quality improvement. Further shape and content will be given to this initiative in the coming months. The main points of our quality policy are:

(1)   Quality improvement is primarily a task and responsibility of management as a whole.

(2)   In order to involve everyone in the Company in quality improvement, management must enable all employees — and not only the employees in the factories — to participate in the preparation, implementation and evaluation of activities.

(3)   Quality improvement must be tackled and followed up systematically and in a planned manner. This applies to every part of our organization.

(4)   Quality improvement must be a continuous process.

(5)   Our organization must concentrate more than ever on its customers and users, both outside and inside the Company.

(6)   The performance of our competitors must be known to all relevant units.

(7)   Important suppliers will have to be more closely involved in our quality policy. This relates to both external and internal suppliers of goods as well as of resources and services.

(8)   Widespread attention will be given to education and training. Existing education and training activities will be assessed, especially as regards their contribution to the quality policy.

(9)   Publicity must be given to this quality policy in every part of the Company in such a way that everyone can understand it. All available methods and media will be used for internal and external promotion and for communication.

(10)   Reporting on the progress of the implementation of the policy will be a permanent point on the agenda in Review meetings. The Quality Steering Group, under the direction of the Board of Management, together with the Corporate Quality Bureau will provide support and co-ordination at the corporate level.'

One of the implications of such a policy statement is that quality-control inspectors tend to disappear from the scheme of things; instead a Corporate Quality Bureau is created whose function is to help everyone become part of this effort of intercommunication. Each person or group of people, whether in the design, production, marketing or servicing section, is expected to work out — often in 'quality circles'* — their own job specification. On the one hand they

---

* Quality circles were introduced in Japan in 1962, but it was not until the late 1970s that they were first set up here and in the USA. They have been defined in Japan as 'a small voluntary group formed to perform quality-control activities within the workshop to which they belong. This small group, with every member participating, meets regularly as part of the company-wide quality control activities and helps with the self-development and mutual development of those taking part'.

must ensure that they are receiving work and information that enables them to do their job properly and, on the other hand, they must be sure that output meets the needs of the next group. It is the task of the quality personnel to assist everyone in this task and in devising quality measures by means of which their own performance, and that of others, can be assessed.

But this is only the beginning. By emphasizing the importance of continuous improvement, quality assurance is given a dynamism which ensures that the quest for quality is not just a once-only activity. Quality assurance accepts that everyone in an enterprise is just as likely to be able to suggest how best to do their job as is an expert planner, engineer or manager. Experience can be just as important as training. Staff are told as much as possible about the running of the company and are given every opportunity to exert control over their own working practices.

This emphasis on the value of everyone's judgement has led to the creation of a variety of methods for generating corporate ideas, including quality circles, suggestion boxes and competitions, but especially recognition of those whose good ideas which are implemented in practice. The form of recognition can take many forms – from cash hand-outs to company support of favourite charities, etc., and is best left to the choice of the individuals or teams involved.

It is this kind of commitment, from top to bottom of a company, which is now seen as essential to achieve the kind of quality that can compete in the modern world. As Sir Geoffrey Chandler put it in a lecture to the Royal Society of Arts in 1987; the philosophy of quality assurance is 'excellence of product, excellence of treatment of employee and customer'. However it is not a simple matter to introduce such systems into a company, especially when Taylorism has been the current practice for some time. Nevertheless, it is coming to be recognized as the preferred way to increase quality as well as increase pay and reduce costs and wastage.

■ Summary and conclusion

The key concepts concerning quality and quality assurance are:

Quality is defined as: 'The totality of features and characteristics of a product or service that bear on its ability to satisfy stated aims or implied needs', or, more briefly, 'Quality = fitness for purpose'; where the customers indicate the 'purpose' and the provider ensures the 'fitness'.

Implicit in the ability to achieve 'fitness' is, of course, the knowledge and skill of the producers to match the characteristics of the product or system to the stated 'purposes'.

Quality assurance is: 'All those planned and systematic actions necessary to provide adequate confidence that a product or service will satisfy given requirements for quality'; to which the following note is added: 'Unless given requirements fully reflect the needs of the user, quality assurance will not be complete'.

There are two main ways of achieving quality assurance: one is by implementing BS5750 'Quality systems', which sets out a number of actions which need to be taken if minimum standards of quality are to be achieved; the other is through Total Quality Management which involves everyone concerned in the enterprise in the continuous pursuit of quality and its improvement.

These definitions and procedures have been developed with industry in mind, where the *expertise* to achieve quality already exists. Quality assurance is then only concerned with ensuring that this quality is consistently delivered. However attempts are now being made to adapt quality assurance procedures to other enterprises, such as health care and education. At the moment the pursuit of quality in education, both in schools and in higher education, is based essentially on Taylorism — the use of inspectors whose expertise is scarcely more firmly based than that of those they are inspecting. Indeed, 'peer review' or inspection by HMIs is generally regarded as the best way of encouraging 'best practice' — as the inspectors or 'auditors' perceive it. There is little attempt so far to apply the excellent research (e.g. Entwistle and Hounsdel, 1976) that has been done on how to match methods to different stated aims, as is continually done in industry. Only when this expertise in the teaching profession is as widespread as it is in many other professions will genuine quality be possible and quality assurance procedures become successful.

■ References

British Standards Institution (1987) *BS4778: Quality Vocabulary*.
British Standards Institution (1987) *BS5750: Quality Systems*.
Entwistle, N. and Hounsdel, D. (1976) *How Students Learn*. Lancaster University.

2.6

Control

T394 course team

The purpose of this [article] is to show the breadth of application of control ideas. The first part, using examples from different spheres, distils the underlying principles and strategies of control. The [article] continues by examining the role and the duties of the control engineer in designing and analysing a particular control system.

■ The principles of control

Control ideas are part of our everyday lives. The word 'control' is used by engineers, accountants, generals, production engineers, store managers and many other people. Although these people come from different fields, handle different commodities, have different aims and use different tools, they are all involved in exercising control in their various areas and their notions of control have an underlying similarity.

Control engineers are also involved in the use of control, their work is concerned with the control of engineering systems: paper-making machines, guided missiles, glass-making, satellite positioning, bread and biscuit making machines, oil refineries, power stations, diesel engines, chart recorders and many other machines and processes. The bias of control engineers towards engineering systems is not the only factor which distinguishes them from other users of control ideas. Control engineers would not personally control an oil refinery or a guided missile, they are concerned with machines that control. Control engineers design and implement automatic control systems.

Before anyone can exercise control over any kind of system he requires an objective. Every controller tries to achieve a certain goal through his actions. The accountant's objective is to balance his accounts; the production manager has to ensure that his product is produced on time; the general needs to defeat the enemy; the store manager has to ensure that goods are available to his customers.

133

However, an objective, although necessary, is not enough. The controller must have freedom to make changes to the system under control. A general gives orders to his men; a store manager requests new stocks from his suppliers. The control engineer's automatic controllers must also be able to make adjustments to the systems they control. All controllers initiate actions, and their actions result in changes that achieve the controllers' objectives.

Unfortunately, complete freedom of action is seldom available to a controller, because there are often constraints that limit the range of possible control actions. The general's men cannot be expected to continue without sleep; the accountant cannot stop wages being paid; the production manager cannot expect new machines to be available instantly. If control is to be achieved, account must be taken of all possible constraints.

The nature of the constraints facing a given controller may be such that the original objective is not immediately attainable. Realistic control objectives are therefore formulated that specify a goal in terms of a measure of performance over a period of time. The accountant must balance his books at the end of the year; the general is given a number of days to gain ground.

> If control of a system is to be possible, the controller must have the opportunity to take actions that make adjustments and cause changes to the system. If control is to be achieved, the controller must take into account any constraints that may act on the system that is to be controlled.

A controller cannot change the present, he can only take actions that have future effects. It is therefore essential, if control is to be achieved, for a controller to be able to predict the consequences of possible actions so that he may select which actions should be taken and decide when to take them. Prediction requires a model based on past experience. The decision to make a model may not even be a conscious one. The model may be naïvely conceived, but it is essential to the achievement of control.

The model that the store manager uses might include a notion of the rate at which stocks are used and the time it takes for a supplier to deliver fresh goods. The model that the general uses might include maps of the terrain, a conviction that his men will always obey his orders, a knowledge of the capabilities of his weaponry and the results of studies of earlier battles. The production manager relies on data on the capability of his labour force and his machines, on the maintenance requirements of his machines and on a knowledge of the materials that are used in his product.

There are many ways in which control can be achieved. From a model, perhaps based on an experience of similar systems, and from a knowledge of objectives and constraints, the control engineer, the general, the accountant can each create a strategy – a plan of action. Again, this may not be a conscious act, but in each case a requirement exists for a framework in which decisions can be made and appropriate actions chosen in order to secure the necessary quality of control. This strategy may govern the orders that the general gives, or it may govern the way in which the control engineer connects together the equipment that performs the automatic control.

The first strategy I will examine requires a plan or schedule based only on a knowledge of the likely outcome of control actions. Figure 1 illustrates the principle. The production manager is given a requirement in terms of the level of production. He then uses his schedule, which is based on the past performance of the factory, to choose the actions that he initiates. The model incorporated in the plan can predict the consequences of control actions. The model can also be used to search for the control actions necessary to achieve a required outcome. These control actions can then be applied to the controlled system to achieve the objectives. This strategy depends on complete faith in the accuracy of the model that describes the effect of control actions on the controlled system. It also assumes that outside influences are negligible. No checks are made on the outcome.

The accountant may have reason to believe that when he sets spending limits they will not be exceeded. The general may assume when he issues an order it will be carried out. It is unlikely in either case that there will be no overall checks performed to ensure that the actions taken are effective. However, in small ways, control actions will be taken the effects of which will not be checked and for which allowances for error will not be made. In such cases, orders are given and are subsequently carried out. This kind of strategy is called *open-loop control*.

If changes in outside circumstances are known to have an influence on a system that is to be controlled, it would be foolish for the controller to ignore them. It may be possible to remove or reduce such influences, but, if this is not adequate, another strategy is required. The model must be improved and extended so that the effects of outside influences or disturbances can be predicted. Actions that counteract the effect of the disturbances must then be worked out and executed. Figure 2 shows how this strategy operates. Outside influences (such as a shortage of raw materials) affect production, but through the plan correcting actions are taken (for example, the ordering of an alternative material).

A policy of counteracting the effect of outside influences where they are significant is clearly better than disregarding them. However, in order for this strategy to be implemented, the influences must be quantified and the plan or schedule must be accurately extended. For this strategy, where outside influences are monitored and control actions chosen to counteract them before

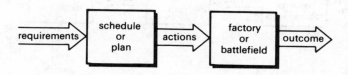

Figure 1 A simple control strategy.

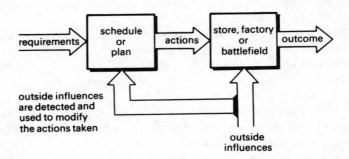

Figure 2   A strategy that takes external influences into account.

they cause any serious deviation from the goal, the term *feedforward control* is used.

Feedforward can be a practical solution if there are a few disturbances that can be conveniently measured. However, it can be an impracticable solution if there are too many disturbances or influences to cope with, or if unforeseen disturbances occur. In such a case, there is one other strategy that can be adopted that again involves measurement. Instead of observing the outside influences, the effect of these influences is measured as a deviation of the whole system from the specified requirements. When the deviation occurs, correcting actions are worked out and applied. The principle of this strategy is shown in Figure 3. Outside influences are still present and will affect the outcome. However, if the outcome does not meet the requirements this will be detected, and actions will be taken that will attempt to restore the outcome as required.

This strategy involves the measurement and comparison of quantities related to the objectives set for the controller. The store manager has to maintain stocks; he can measure those stocks by counting them, and if they are inadequate he can place orders. The accountant can monitor the bank account,

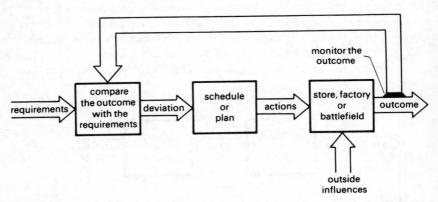

Figure 3   Measuring the effect of disturbances and countering that effect.

calculate his company's actual rate of expenditure and apply any necessary budgetary constraints.

The term *feedback control* is used to describe this strategy. The control actions create an outcome which is measured, and the measured value is fed back to compute a new control action when deviations from the objective occur. There is said to be a *loop* because the control action results in an effect which in turn is used to calculate a new control action. *Closed-loop control* is a term that is sometimes used in place of 'feedback control'.

The feedback control strategy does not require an accurate model, because errors in control actions resulting from an incorrect knowledge of the controlled system are reduced in the same way as the effects of disturbances. However, corrections can only be made after a deviation has occurred and has been detected.

Often there are delays between a control action being taken and its effect becoming apparent. If the model has not recognized these delays, an immediate effect is expected from a control action. Because of the delays, no immediate response will be detected and this apparent lack of response will cause the controller to increase the control action. Eventually, the result of the increased control action will be seen as an excessive outcome. The attempt to compensate for this excessive outcome will lead to excessive control action being taken in the opposite direction. The whole cycle will continue, and the outcome will experience fluctuations. This is clearly undesirable. The accountant would be castigated if there were fluctuations from heavy borrowing to excessive surpluses. The store manager would be unpopular if, by ignoring the delay between ordering goods and those goods being delivered, his strategy caused severe stock fluctuations. The possibility of this fluctuating outcome resulting from the application of feedback control is always one of the great concerns of the control engineer.

> There are three principal control strategies: open-loop control, feedforward control and feedback control. Open-loop control assumes that the effect of disturbances and variations in the model are negligible. Feedforward control attempts to correct the effect of disturbances by measuring those disturbances and predicting a counteracting control action. Feedback control corrects the effect of disturbances by first measuring their effect and then calculating a correcting control action.

■ A control system

Control engineers design automatic controllers. In designing a control system the control engineer has first to be clear about what the automatic control is to achieve. Objectives are often simply stated, but are less easily interpreted. With a missile the objective is clear: to destroy the target. The management of a rice-pudding factory may wish to add automatic control to improve profitability, to increase production, to reduce the labour force or to meet the requirements of new legislation. The possible objectives are manifold.

Secondly, a control engineer must reduce these objectives to a form that can be interpreted in terms of the plant or equipment. To do this it is essential that he understands the processes or machines, not only how they work but also their economics, associated legislation and operating personnel.

The milling of flour can serve as an example. The cost of imported grain continues to rise and attempts are being made to increase the amount of ground flour that can be separated from the husks. The objective is to increase the profitability of grinding grain. From the experience of centuries of millers, it is known that the moisture level in the grain affects the yield, and so grain is moistened before it is mixed with other grains to make up the grist for the mill. There is a level of moisture that maximizes the yield from the grain. The objective of increasing the profitability of milling can be reduced in part to an objective which requires a level of moisture to be maintained, provided that the cost of the controlling is included when assessing whether the new proposals will increase profitability.

> The first job of the control engineer is to determine the objectives of his control systems, and to express those objectives in a way which enables him to apply the techniques of control theory to lead to a practicable design.

For the grain moisture control system, the moisture of the grain can be adjusted by adding water. The flow of water could come from a tap as shown in Figure 4. If the tap can be adjusted automatically, then it can be used in an automatic control system. Devices which enable an automatic controller to make adjustments to the system under control are often called *control elements*. It is essential to have a control element if actions are to be taken to achieve control.

There is a maximum rate at which water can be added to grain from a particular tap; and moisture can only be added from a tap, it cannot be taken away. This is a constraint on the possible actions of the control system.

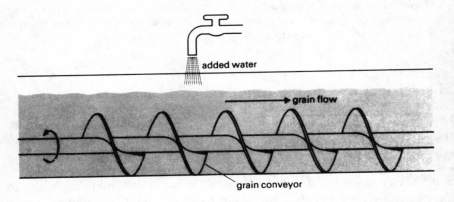

Figure 4 Adding water to the grain.

Having an objective, an ability to take automatic control actions and a knowledge of the constraints, the control engineer next needs a model of the equipment or process to be controlled. The model may be crude; perhaps the simplest model the control engineer needs is a description of what changes can affect the quantity he is to control. Adding water will increase the moisture content of the grain.

Increasing the sophistication of the model gives the engineer an opportunity to improve the quality of control. This does not mean that the control engineer searches for the most complicated model. The model should be simple so that he can use a simple design procedure; but it should be adequate to enable controllers to be designed to meet the specification required of the controlled system.

For the grain-moisture control-system design, the model can be improved by relating the moisture in the grain to the flow of added water. If it is assumed that the grain flow is constant and that the initial moisture content of the grain is fixed, then the steady flow of additional water that will produce the desired moisture can be estimated from the model. A second part of the model might relate the tap setting to water flow. For a required water flow it would then be possible to work out the corresponding tap setting. The tap can be set and the correct amount of water will be added to achieve the required moisture content.

Figure 5 represents a proposed grain moisture control system [. . .] Could feedforward improve the control of grain moisture if variations of incoming moisture content, grain flow and water pressure were known to exist? What are the implications for the controlling equipment if this is to be done automatically?

If all these quantities are measured, the results of the measurements could be used in calculating adjustments to the tap setting which ensure that the ultimate grain moisture content is as required. To be able to do this automatically the measurements of grain flow, incoming grain moisture and water pressure must be made automatically. Given a measurement for the incoming grain moisture, the controller must work out, automatically, what the tap setting should be to achieve the necessary water flow.

Similarly, it must be capable of calculating the adjustments for changes in grain flow and water pressure. These adjustments can be calculated using the model. Therefore, the controller must do calculations based on the model. Finally, the controller must be able to operate the tap − automatically.

Before feedforward control can be applied, a measurement of the incoming

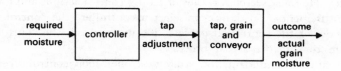

Figure 5 A proposed grain moisture control system.

grain moisture content is needed. The measurement must be automatic, and must be able to operate automatic equipment. A transducer is available which gives an electrical output which indicates the moisture content of grain. Now, through the use of the model, a controller can be devised that adjusts the water flow depending on the transducer output. This gives an adjustment for changing incoming grain moisture, but it gives no adjustment for changes in the flow of grain or for changes in supply water pressure. Transducers could be used to measure water pressure and incoming grain flow as well, but because there are now three parameters involved, the controller needed to implement the control would be quite complex.

In these circumstances the control engineer would try to remove the other causes of disturbances; he might add a header tank to ensure that the water supply pressure is constant and a hopper to try to maintain constant grain flow. However, it may not be possible to maintain a constant grain flow, and there may be other unforeseen disturbances. Feedback control is able to compensate for unforeseen or ignored disturbances, and it can also compensate for errors in the model. To implement feedback control, a measurement of the outcome of the control action is needed. This means that the moisture of the outgoing wet grain has to be measured.

The final control system in use for the control of grain moisture uses a feedforward and a feedback strategy: feedforward to accommodate changes in incoming grain moisture, feedback to accommodate all other variations.

This scheme is shown in Figure 6. There is a transducer to measure incoming grain moisture and a transducer to measure the final grain moisture. The signals from both transducers are combined in the controller to give the necessary automatic valve setting for the correct water flow.

[. . .]

■ Summary

The principles of control are frequently used by many people in the course of their work. In the case of control engineering, control principles are used to design automatic controllers.

In establishing a control scheme, the control engineer must derive a realistic objective, he must identify possible constraints and sources of disturbances. He must search for ways of taking control actions and for ways of taking measurements. To design or select a controller, the control engineer needs a model of the controlled process. He must install the controllers and adjust them to suit the actual process.

There are three fundamental strategies: open-loop control, feedforward control and feedback control. To help design control systems using these strategies there is a body of mathematics referred to as control theory.

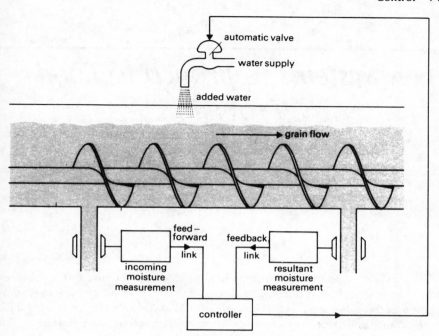

Figure 6 The grain moisture control scheme.

2.7

A Systems Approach to Food Provision

T274 course team

[This article is taken from a course on food production systems, and it examines how to represent these systems. The discussion starts by looking back over some of the aspects and activities of food production that have already been considered in the course and tries to create models that will represent these activities. The purpose at this stage is to better understand the food production systems and to illustrate a systems approach.]

[. . .] Our initial examination of the factors affecting food supply in a small community only addressed one part of the whole situation and there are several other levels at which things need to be examined. There is obviously some sort of international dimension to our concern, with the relationship between the EEC, the USA, USSR and other nations having an important bearing on the production and supply of food. At the other extreme, some of the problems we identified involve the reaction of the individual to very local events, or even to factors, such as the sense of hunger, which are internal to that individual. Although these concerns are all part of the same total system, they often arise from different *subsystems* within it. It would be unrealistic to expect any one model to be perfect for all these levels of concern. We will often need to use different models to represent these subsystems and the relationships between them, but it is nevertheless useful to try to decide what overall system is giving rise to the set of concerns.

If we start at the broadest level of resolution, considering the global balance between supply and consumption of food, we can visualize this situation in terms of two subsystems, a subsystem which supplies food, and a subsystem which expresses a need for food. Note that I have chosen the words to describe the two subsystems with some care, in a way which accords with my particular view of their nature. The first subsystem 'supplies food', whereas the second subsystem 'expresses a need for food'. This distinction is particularly clear if you look at the situation in the Western (or Northern) world. Here, the *production* of food, on farms and in food processing factories, is a clearly separate activity from those giving rise to consumer demand for food. The two sets of activities occur in different geographical locations, and, by and large, involve different sets of people.

The applicability of this model to a peasant household, with its own land, or access to enough land to provide its annual food needs is not quite so obvious. Here, apparently, the two subsystems involve the same people, in the same area. However, if you think about this a bit further, it should be clear that the *activities* represented by the two subsystems are very different in kind. In the first case, the series of activities includes preparing the soil, planting crops, tending and harvesting them, managing livestock and preparing acceptable food from these products. In the other case, the action is expressing a need for food, which even at the level of the independent peasant household involves separate individuals, reacting to a different set of stimuli from those associated with the production activities. Even in this (probably almost fictional) independent peasant household, there is still the central task of balancing these two subsystems. The household has to decide how much, of what crops, to plant, how much time to spend tending those crops, what to do with them when harvested, and so on and so on, in relation to food needs and wants today, tomorrow, and for the rest of the year. Thus, it may be useful to regard the two subsystems as being separate, if only because it highlights the importance of the *links* between them.

There are several distinct forms of linkage between these two subsystems even at the level of the independent household. Thus, there is a *flow* of food materials from the production subsystem to the 'food demanding' subsystem, and there is also some linkage concerned with the transfer of information about food requirements from one to the other. The *processes* involved in making decisions about the allocation of effort to the various activities in both the peasant household or on the wider world scene can also be visualized as another subsystem. This subsystem allows the *exchange of information* and the *transfer of goods* between the first two and, for the time being, I will refer to it as an 'exchange' subsystem. In any commercially oriented society, monetary exchanges are likely to be a major function of this subsystem; in the self-contained household, this aspect may be less important.

[Consider] some of the activities involved in this [exchange] subsystem in the UK. The obvious function is that of passing money around; consumers have to pay for food, producers need to borrow money before harvest, or to invest in new machinery, etc. So, the banking system presumably figures in here. However, you might also have thought of the effects of education and advertising on the expressed demand for food, of the role of food in international relations, and of the other, social roles of food (as in dinner parties, which have very little relation to nutrition!), etc.

Separating out three subsystems in the overall food system does two things: it reminds us that the activity of feeding people has several very different and disparate aspects, and it gives us the bare outline of a model with which to structure our investigation of the food system. Let us have a look at these subsystems to see what goes on in each, how they are linked together and what further knowledge or techniques we need in order to understand and model them.

Since it is both the most familiar to us in everyday life and involves a major element of individual preference, perhaps the simplest subsystem to examine is the one which expresses food needs. At its most basic, we are all familiar with our feelings of hunger, and the resulting actions which we are likely to take to ensure that these feelings are satisfied. But what may not be so obvious is *why* we feel hungry. [. . .] To answer this we need at least to know something about the physiology and biochemistry of our body. [. . .] You are probably also vaguely aware that your actual choice of food is influenced by things other than just how hungry you are. You almost certainly have some pet likes and dislikes in foods, which you would be hard pressed to justify on biochemical grounds.

As a simple example, would you feel at ease eating porridge for your midday meal, or fish and chips for breakfast? Certainly if you were hungry enough you would do this, but it would be unusual, to say the least. This apparently banal example is designed to show that the expressed demand for food, as opposed to the biochemical need for particular chemical components of food, is itself the result of a complex of factors. Some of these, like the British custom of eating toast and marmalade for breakfast but not at other meals, are the reaction of individuals to various externally determined norms or conventions of a particular society. These norms and other similar pressures are perhaps most correctly regarded as being part of the 'exchange' system suggested earlier. If we are to understand why particular dietary patterns, good or bad, arise, we need to examine both the biochemistry of food need and these other, social, factors, to see how they interact. The supply side of our system involves several other generally familiar activities. [. . .] The obvious ones are growing crops and feeding animals, but remember that we are talking about food, not just agricultural produce, so there are also the activities of storing and processing these products and their distribution to consumers.

Although we are all probably vaguely aware of the methods of agriculture and food processing, at least as amateur gardeners or cooks, both of these activities themselves involve many different subsystems. An obvious feature of the relationship between agriculture and food processing, as also between the 'food producing' subsystem and the 'food requiring' subsystem is the *flow of material* between them. In the UK, food 'flows' from farms to factories to shops to people; in our 'peasant' household there is a flow of food from field to cooking to consumer. This notion of the flow of food materials is just one possible relationship that may exist between subsystems in the overall system of interest to us. [. . .] In talking about the 'food requiring' subsystem, I suggested that there was some exchange of information between this and the food producing subsystem, and that there was some *influence* from society at large on the type of food demanded.

I suggested that society at large was part of the rather nebulous 'exchange' subsystem and that this subsystem was involved in regulating the activities of production and consumption. The relationship between this subsystem and the 'food producing' and 'food requiring' subsystems is more subtle than that of the obvious *flow of materials*, but it is none the less important to our overall understanding of food production.

What should be beginning to become apparent is the need to have some method of representing these systems and subsystems, and the relationships between them. Again, we are faced with a situation where there is not a nice, neat linear sequence, but potentially a series of loops and other nonlinearities. I would not be surprised if, in reading the preceding text, you tried to visualize the various subsystems in some pictorial form and, once more, diagrammatic representations are likely to be helpful. However, I deliberately did not use a diagram at this stage because, to be generally useful, we need first to set up some agreed conventions for drawing such diagrams, so that we all mean the same things by a given picture.

In the following sections, we will look at two [kinds of diagram]. You may find that the next few pages seem to be quite heavy going. While diagrams themselves are a very quick and efficient means of communicating information, using text to describe *how* to use them, in an agreed manner, does take some time. However, you will find the end result will be useful, so it is worth spending some time on this topic.

■ Diagrams of systems

You may have wondered why we could not just use the multiple-cause diagrams[1] as a general technique for representing systems. [. . .] In a multiple-cause diagram, what we are doing is to represent the relationship between a number of factors, and some resulting *event*. The lines on the diagram suggest that a particular factor causes, or at least affects, some other factor, and the effects link together to produce the end result in which we are interested. [. . .]

There are two main reasons why such a diagram would not be an appropriate method of representing the relationships between subsystems. Firstly, a multiple-cause diagram shows linkages between single *factors* or *events*, not between systems or subsystems. [. . .] Secondly, the relationship between factors in a multiple-cause diagram (i.e. the concept of causality) is very different from either the more vague idea of 'influence of one subsystem on another' or the very precise notion of a 'flow of material between subsystems' which we have introduced [. . .].

It is clearly impossible to summarize any of the suggested subsystems of our overall food system in terms of a single number on a scale, so we need to find some less restrictive method of representing these subsystems and their relationships. Alternatively, where we know that the relationship between subsystems clearly involves *flows*, we need to have another, different, method of representing this. In this [article], we will use what are called *influence diagrams* for the first of these cases, and *flow–block diagrams* for the second.

☐   Systems maps and influence diagrams

An underlying concept of all the preceding discussion is that we can define some overall system, which relates to our topic of concern and within which there may well be some relevant subsystems. One way to capture these notions in diagrammatic form is to draw what is called a *systems map*. This is a rather less formal version of the mathematician's Venn diagram, where items which share some characteristic are shown within a boundary line. Items which do not have the defined character are then outside the boundary. The elements of this form of representation are shown in Figure 1.

In Figure 1, you can see that we represent the system that interests us as an irregular shape, within which the subsystems are represented by smaller shapes. Outside the main boundary there may be other systems, represented by further enclosed shapes. The line round the main shape is the *system boundary*, and it separates those activities, items, processes, etc., you feel are an integral part of the system from those outside the boundary which are not part of it, although they may affect what goes on in the system.

To see how this works and is relevant, let us go back to the example we have been using so far, that of a completely generalized overall food system. In this system, I suggested that we can outline three subsystems, involving production, food requirement and a subsystem which provides for the monetary or other exchanges involved in regulating or managing the other two subsystems. We can represent this on a systems map using Figure 2, where the three subsystems are shown within the overall boundary of the 'food system'.

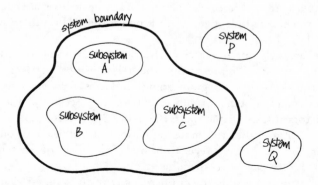

Figure 1   A stylized systems map.

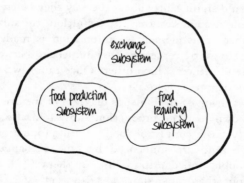

Figure 2 A systems map of the generalized 'food system'.

We have still not defined very clearly what goes into the third, 'exchange' subsystem. One very important effect of drawing systems diagrams is that the activity forces us to make explicit some of our assumptions about the subsystems we have chosen. If we think solely about the food system of the UK, then we can probably agree that the food-related activities of farmers and food processors are the major part of the production blob in Figure 2, and the activities of consumers form the food-requiring blob. The third, the exchange system, does not map so easily onto any recognizable group of people, but is defined more by the activities carried out in that subsystem. These activities are those of co-ordinating the production and consumption activities and relating them to other aspects of life. So we noted earlier that within this blob we would include the social pressures to conform to certain social norms and the financial systems that are involved in most food transactions.

Still thinking only of the UK, what other *activities* would you include in the exchange subsystem? One obvious one is Government legislation; think [about] the mandatory labelling standards applied by Government to food products. Another is the transfer of *information*, about what food is available, and what are its properties, perceived or actual, via advertising or education. A third is probably Government policy, which defines much of the 'climate' within which basically financial transactions take place.

Even with the accompanying text, Figure 2, like many systems maps, does not add a great deal to a simple list of what we believe to be the important subsystems of our overall system. What it does do, however, is give us the basis for our influence diagram, which we can then use to show some more of the important *relationships* between the subsystems. The map showed only one aspect of these relationships, namely the fact that we regard all three subsystems as belonging to one overall food system. This in itself is actually not a trivial statement; all too often the food system has been analysed as if it comprised only the activities of production, or of consumption, or of economic interaction. Such analyses explicitly assume that there is no interaction between these subsystems which could significantly affect the way they behave. Putting them together as parts of one system says that, by definition, the subsystems interact and affect one another. This element of interaction is made explicit in the diagram by adding lines to represent the influences between subsystems, specifying which one influences which other. Once more, we do this using an arrow, this time to denote *influence*. [. . .]

Figure 3 is my version of an influence diagram of the overall food system. Note that I have shown that the production subsystem affects the requiring subsystem directly (via the supply of food). Both production and requiring subsystems influence, and are influenced by, the exchange subsystem, as indicated by the double ended arrows linking them.

It seems obvious that the requiring subsystem affects the production subsystem. Why is the arrow linking them not double headed? The reason for this lies in the interpretation which I have placed on the exchange system, in the UK context. It is arguable that while production affects consumption directly, via the supply of food, the reverse effect is different in kind, comprising market demands, information, etc., all of which are transmitted via the exchange system.

Do you think that this distinction is valid for all types of food systems? For example, what about that represented by our hypothetical peasant family? Probably not; consider the case of a peasant family which is only just at subsistence level. Here, the supply of food to the consumers will directly affect their ability to work at growing more food. In this situation, it is probably more correct to show the direct link between these two systems as also being two-way.

If you think about what we have so far included in our overall food system, you will probably realize that there are several other sets of interlinked activities (*systems*) which have an influence on the food system. One obvious one is the world weather system; should we put this inside the system boundary or not? My answer is no, since while it certainly affects the food system, the food

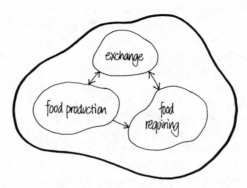

Figure 3   An influence diagram of the 'food system'.

system itself has only a minute effect on weather systems. This is a very useful criterion for deciding whether or not some set of activities lies within or without the system of interest.

What other systems should we bear in mind, outside the general food system, which nevertheless have some influence on it? The most useful suggestions are probably the rest of the 'world trading system' and 'other industries'. These two systems probably have their main effects via the exchange system, although [. . .] the industries which supply machines and agrochemicals have a direct effect on the production subsystem. We indicate this by drawing the influence arrows from the systems outside the boundary to the relevant subsystem within. One other system which may be relevant is a 'health-care' system; I initially thought this should be represented, but outside the boundary. However, on thinking about this further, I decided that the effects of nutrition on health are so profound that it was probably more relevant to include this within my food system. Depending on the purpose for which the diagram was being drawn, and the exact definition of the 'health-care' system, this may or may not be appropriate.

My next version of an influence diagram of a generalized 'food production and consumption system for a single region' forms Figure 4.

You may be wondering why I have had these diagrams drawn with irregular shapes for the boundaries. This is not pure caprice on my part, but is a deliberate device to indicate that this is only my *interpretation* of the food system. If I had had it drawn very carefully and geometrically it would have

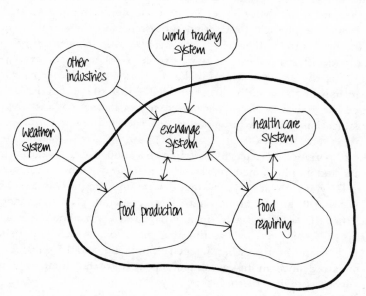

Figure 4   An expanded influence diagram of the food system of a defined, but unnamed, region.

given the diagram an authority which it does not deserve. There is also a danger that in laying it out in an aesthetically pleasing way, I might have been tempted either to ignore certain links, or to rearrange things in such a way that their meaning is changed. It is often useful to indicate that particular subsystems are closely related by drawing them close together, and such ideas should take precedence over artistic nicety! When drawing such diagrams yourself, you do not have to avoid using compasses and a ruler, but, if you do use them, be careful not to fall into the traps I have just indicated.

In Figure 4 I have indicated the subsystems as shapes within the system boundary and other systems which affect the food system outside the boundary. Note that while in this case, we are considering a defined, though hypothetical, geographical area, the system boundary does *not* necessarily correspond to any geographical boundary. The system boundary represents the *intellectual* and *functional* distinction between those items which are directly involved in, and those which influence but are not directly involved in, the situation. One other aspect of these diagrams which sometimes causes problems is overlap between subsystems. After all, you may say, the same people will probably be members of more than one of the subsystems listed here, and so, should this not be indicated by overlapping the subsystems to show those common members? This is what happens in the mathematicians' Venn diagram: however, what we are trying to do in this influence diagram is to separate out important *functional subsystems* and the relationships between them. The distinction between the subsystems lies in their function, not in the people who happen to be members of them. Indeed, most of the subsystems, such as the production system, contain elements other than people. So, for example, in the case we are looking at here, while the physical production subsystem and the social organization subsystem may both involve the same people, they concern very different activities, and different aspects of those people, so they do not overlap in the diagram. It is the *activities* which make up these two subsystems which are important, not the actual individuals involved. So, in the production subsystem we include the activities of *growing, processing and storing* food, whereas in the social organization system we are mainly concerned with *communication* of ideas, norms, needs, etc.

The importance of visual models like this influence diagram is that they are a simple and graphic way of representing *your views* of what is important in a situation. They are a relatively informal device, and the individual subsystems are not necessarily at a commensurate level of detail. This does not mean, however, that they can be drawn up in a haphazard way. The diagram should summarize the major factors, within that situation, which you believe are important in determining what happens, and you should ensure that the subsystems are defined in any accompanying text so that their properties are clear to anyone studying them.

[. . .]

☐   **Flow–block diagrams**

The basic form of these diagrams arose mainly in the chemical process industries, where many of the activities concerned are *flows* of materials between different processing operations. So, for example, liquid raw materials may be stored in a series of tanks, flow to a mixing tank, then flow on to some other process tank, finally being held in store until dispatched to customers. Conceptually, it is very easy to represent such activities using a series of lines to represent the flow of materials between (usually rectangular) *blocks*, with the blocks representing the tanks or other plant in which the various operations occur. Figure 5 is a simple example of such a diagram, and Figure 6 is a real example taken, incidentally, not from the chemical industry but from a food processing plant. Conceptually, the link between such diagrams and the industrial processes they represent is clear and simple. However, researchers in other subjects and other industries found that there were often ways in which they could also make use of this type of model to represent their own topics of interest. For biologists, such models were a powerful means of representing the functional behaviour of *ecosystems*, where the flows might be plant nutrients, chemical compounds, energy or a number of other items, between different sorts of organisms. Each type of organism could be represented by a different *block*, and the *rates* of flow between these different blocks are often a major determinant of the overall behaviour of the whole system. In order to indicate that the rates of flow are of importance, each rate is often labelled, using something like the valve or flag symbol shown in Figure 7, which is a highly stylized and simplified diagram of a typical ecosystem. You may wonder how there can be a 'flow' of material from nowhere into the block labelled 'plants' — think about this for a moment, to see what it represents.

You are probably aware that plants grow by taking in carbon dioxide, water and nutrients from their surroundings, and assembling these into their structures. We could have represented this by having a further block as the 'environment' from which these materials were obtained. However, for many purposes, the removal of these materials from the environment has a negligible effect on the amounts there, so it is not worth considering the environment as a block in the system. In some conventions for drawing such diagrams, these 'sources' are represented as 'clouds', to indicate that they are effectively inexhaustible (on some defined timescale!). [. . .]

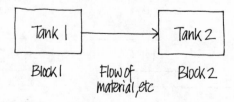

Figure 5   The elements of a flow–block diagram.

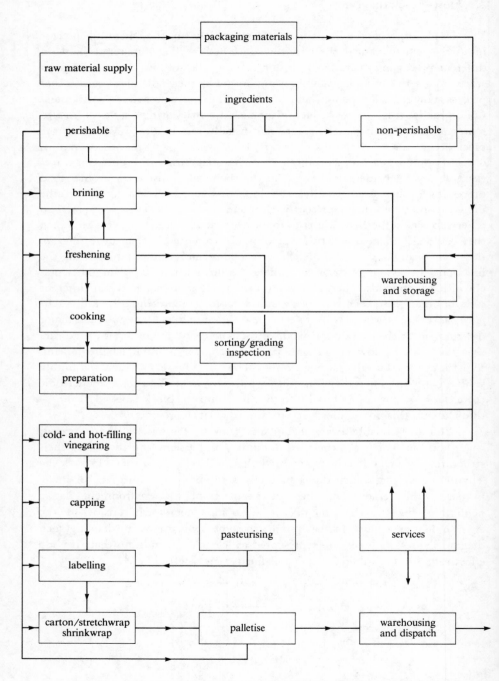

Figure 6 Flow–block diagram of a food processing plant.

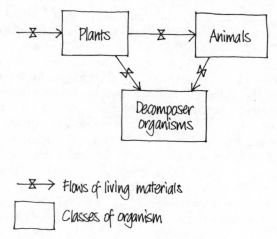

Figure 7 A stylized ecosystem represented as a flow–block diagram.

[How might such a diagram of an ecosystem] be relevant to food production? If you think about the normal processes going on 'in nature', they include the growth of plants, the eating of these plants by animals and the eating of animals by one another. In each of these cases, you can see how the process can be represented as a flow (of plant material, or of animal tissue) between the appropriate blocks. A similar set of processes is involved in our production and consumption of food.

Figure 7 is a diagram of what ecologists call a *food chain*, and a very similar diagram can be used to represent the 'human food chain' [. . .].

[What are] some possible candidates as flows in a diagram of the human food supply system? Depending on your particular interests and experience, you might have suggested just food as the flow in the system, or you might have considered a finer distinction such as the biochemical one between proteins and carbohydrates. Alternatively, you might have thought that it was more appropriate to think in terms of individual commodities, such as meat, flour, vegetables, etc.

Obviously, there are several possible candidates for relevant flows, and to a large extent, it depends on the *purpose* for which you are drawing the diagram. For the present, we only need a relatively crude model, so we can work on the basis that there is only a single commodity, called 'food', encompassing all the variety of items which might come under that heading.

Suggest some of the *blocks* (in the sense that we have been using the term here) that you might include in a flow–block diagram of the human food supply system. Again there is a range of possible answers. You might have chosen items like farms, food processing factories and shops, or you might have chosen different countries, with import and export flows of food between them. If you were involved in agriculture, you might well have separated out crops, livestock, storage facilities, agricultural merchants or some other series of appropriate blocks, all linked by flows of this general commodity, food.

You may find it rather difficult to see what all the possible candidates as blocks have in common, since in some cases they are easily identifiable with the basic concept of a tank in a chemical plant, but things like 'crops' may seem to be conceptually rather different. Probably the easiest way to consider whether it is sensible to include something as a block, is to think of the diagram as representing a 'snapshot' of the system in some instant in time. If you were to look at the real system at such an instant, then you would, in theory at least, be able to measure how much of the flowing commodity was present in each block, and this would provide a useful description of the system. However, even this degree of detail only gives a partial description of the system, since it gives no indication of the dynamic, time-varying aspects. For a more complete description you also need to know how much of the commodity actually flowed from one block to another *in some defined period of time*. That is, we also want to specify the *rates* of flow in the diagram. In some cases, we may be able to measure these rates directly, or at least we may be able to estimate their relative magnitudes. However, it may still help us to understand how a system operates if we can just indicate what it is that determines these rates, even if we cannot measure or calculate their exact magnitude.

In a chemical process plant, where there are real flows of liquids along pipes [. . .] there are two major factors [affecting the rates of flow]; one is the opening or closing of the control valves, another is the physical or spatial relationship between the flows themselves.

So, in a system comprising a sealed mixing tank with three inflows and a single outflow, if the three inflows are controlled by valves, the outflow is automatically determined as the sum of the three inflows, unless the tank is either emptying or about to explode! This is shown in Figure 8(a). Notice that I have added an additional form of line (the dotted one) to indicate that the rates of flow are interlinked. The dotted lines indicate *controlling influences* which determine the rates of flow. In this case, we suggested that the rates of inflow were determined by some external control agent, who opened or closed the valves on the incoming pipes. In the convention that we are going to use [. . .], these controlling influences are indicated by circular boxes as used in Figure 8(b). We now have a very powerful means of representing a variety of systems which can be conceptualized as comprising flows of some measurable factor, between a set of stages. These stages may be places, activities, people or whatever. Several aspects of the food system can be visualized in this way.

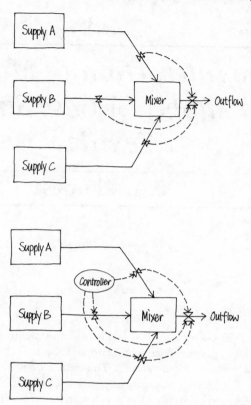

Figure 8 Flow–block diagram of the flows through a hypothetical chemical mixing tank.

Editor's Note

(1) As the text goes on to say, these diagrams link a number of factors with some event. So, for example, it will be possible to link a number of environmental, dietary and genetic factors to death from heart disease. The relationship between the factors and the event can be represented diagrammatically. The 'event', death from heart disease, does not have one single cause (such as excessive smoking) but many causes, hence the name 'multiple-cause diagram'.

2.8

The Interaction of Markets and Supply: A Case Study of Textiles

E.A. Rhodes

■ Overview

During the last 30 years or so, low levels of economic growth and deteriorating international trade performance in Britain and the USA have been accompanied by a search for possible explanations. The nature of technological performance has been the subject of increasingly intense investigation. In particular, attention has focused on innovation performance, primarily on the rate at which new products and new production processes are successfully developed. The historical evidence shows that rates of innovation vary considerably over time, and that there are differences in innovative performance, both between firms and at the aggregate level of national economies. These variations are only partly a matter of major innovations, since small-scale incremental changes within firms can lead cumulatively to major advances in technological performance. Much of the success of Japanese firms has derived from a focus on incremental innovation to achieve continuous improvements in both product design and specification and in methods of manufacture and distribution.

The growing body of research in this area has identified a very wide range of factors that are important in innovation and in other aspects of technological performance. These include such diverse phenomena as: spending on R & D and on technology related education and training – nationally and by companies; the characteristics of company strategies; regimes of investment appraisal; and the many aspects of work and those which are related to product market conditions (see Figure 1). In this simplified model, innovations may be seen as following from some combination of 'technological push' and 'demand pull'. In the first of these, emphasis is placed on the impetus given by advances in scientific or technological knowledge. These are seen as impelling the development of new production processes or new products – sometimes in

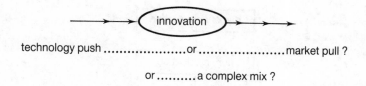

Figure 1 A model for innovation as a combination of 'technological push' and 'market pull'.

completely new markets, as in the examples of the vacuum cleaner in the late 19th century, and the personal computer in the late 20th century. In the case of demand – or market – pull, emphasis is placed on the dynamics of product markets which, in appropriate conditions, such as those where demand is rising and gaps in provision or unsatisfied needs are perceived, may 'pull' new developments through by stimulating innovative activity.

There is space here only to consider a few aspects of the complex interaction between markets and technology. The rest of the chapter will establish some of the salient market conditions in one sector – textiles – focusing on clothing production and retailing in particular. At this end of the production chain, the sector is 'close to textbook conditions for "perfect competition"',[1] and the interactions between these conditions and technological change are particularly clear. (Article 3.7 examines some of the technological characteristics of the sector.)

■ A global production system

External perceptions of the textile sector tend to be of a simple, rather old-fashioned and basic production–distribution chain which, like other 'traditional' industrial sectors, has long been on the downward path of decline. But closer examination shows the sector to be surprisingly complex. Even when viewed at the elementary level of the stages of the production chain which are shown in Figure 2, the route from raw materials production through to end products has more stages than most other sectors. This is partly a matter of the nature of textile products which, from fibre production onwards, generally require a considerable amount of mechanical and chemical processing. In recent years, these processes have undergone a number of changes in order to meet increasingly rigorous standards of performance which are required in an expanding range of product markets. In general, changes in production methods have to allow further cost reductions, greater product diversity, and higher standards of specification and quality. Growing complexity in products has been accompanied by a growing dynamism as the sector as a whole shifts from a craft base to one that is more knowledge and capital intensive at all stages. During the last 50 years or so, this change has also been reflected in the growing role of the textiles related sector of the chemical industry.[2]

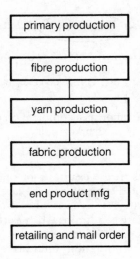

Figure 2   Basic stages in the textile chain.

The extent of complexity in the textile chain is illustrated in Figure 3 which represents the main components within the chain and indicates some of the main production flows. As can be seen, a substantial (and expanding) part of production is directed towards other industrial sectors, and to the expanding 'geotextiles' market in which textile products are used for purposes such as the protection of coastal and river areas against erosion, and in a range of civil engineering functions (for instance, in the foundations of new roads and rail lines). Despite appearances, the figure simplifies the structural complexity of the chain. For instance, it does not indicate the high levels of specialization that are to be found within and around each of the component areas. Specialist firms operate in product areas defined by such factors as fibre type, the category of end product, and the level of the consumer market at which products are aimed. It is important to recognize that the flows in the chain are not solely downwards. While this is the case for material flows, there is an equally important upward flow of design, production, and other information. In this, the role of the retail sector (in clothing and household textiles) is unusually direct since some larger retail organizations have direct roles in design and in the technical elements of product specification.[3]

Figure 3 also needs to be viewed in the context of the global functioning of the textile sector. In the UK, as in other industrialized countries, consumer and other markets are supplied from a manufacturing system which is organized on a global scale, primarily by large manufacturing and retailing companies. Thus, processing in the stages identified in Figure 2 may successively be undertaken in a number of widely dispersed locations. Designs undertaken in the USA may arrive in the shops after, say, production of manufactured fibres in

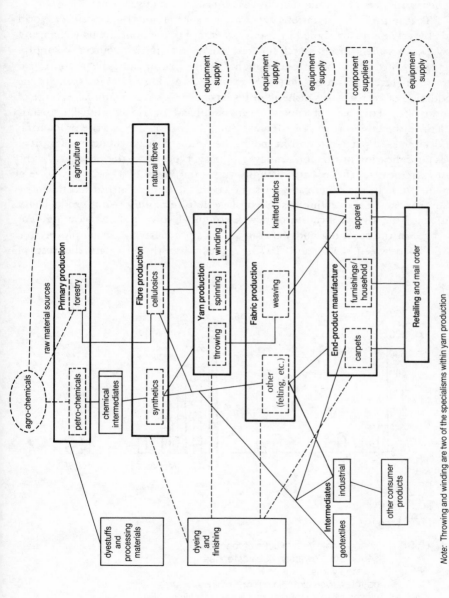

Note: Throwing and winding are two of the specialisms within yarn production

Figure 3   Main components of the textile chain: solid lines indicate downward product flow and interactive design, innovation and information flows; the dotted lines indicate inputs and influences at the various stages of the textile chain.

South Korea, yarn production and fabric weaving in Taiwan, and the cutting of garment components in Hong Kong followed by assembly into finished garments in one of several South East Asian locations.

Textile products of various types account for about six per cent of world trade by value, and are traded in a series of complex and continuously changing networks in which buyers search for fresh advantages in new sources of supply. The emergence of global production has been accompanied by a continued increase in levels of production. For example, the long-term growth in the production of cotton and synthetic fibres has been close to an annual average of around 3%. This has, of course, been paralleled by rising global demand, evident in the UK in rising retail sales. But, in contrast, UK textile production has declined – a pattern found in most other industrialized countries. Figure 4 gives an indication of this for a number of basic types of textile product. In the most extreme case, that of cotton singles yarns, output in 1989 was only 20% of that in 1969. In the other categories shown in the figure, output has fallen by 50% or more.[4] This decline has taken place despite rapidly rising productivity resulting from the application of new methods of spinning and weaving. For instance, in the spinning and weaving sectors, productivity per employee rose by 80% between 1963 and 1978, compared with 54% in the rest of

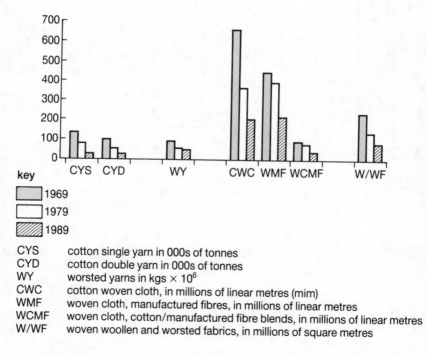

Figure 4   UK textile output 1969–1989, selected items.

manufacturing.[5] The impact on employment levels in the sector has been correspondingly magnified. In 1950, about one million people were employed in textile manufacture from fibre production through to weaving and fabric finishing, while a further three quarters of a million were employed in clothing manufacture. By 1970, the numbers of employees had fallen to 680 000 in textiles and 470 000 in clothing; and by 1990 to 220 000 and 300 000 respectively.[6]

However, despite this decline, the textile sector remains an important part of the UK economy. In terms of value of output, the production components constitute the fifth largest industrial sector, and employed some 500 000 people in the early 1990s. In the retail sector, textile products account for about 16% of all retail sales. Furthermore, current technological and organizational changes may have the potential to stem, and in some cases, to reverse the decline in production (see Article 3.7).

■ Changes in market conditions

The major part of textile retailing is concerned with clothing, which in 1988 had a turnover of £15 billion out of the total £115 billion for the UK retail sector as a whole. The other major areas are household textiles, including soft furnishings, and rugs and carpets – both of which had a turnover of around £2 billion in 1988.[7] In terms of their impact on product development and production technology, the four salient characteristics of the market for clothing products are the level of retail concentration, the maturity of the market, a diversity of sources of supply, and the role of fashion. Each of these is considered below.

In the retail industries of the industrialized countries, there is a general trend towards concentration of sales among large retailing chains and buying groups. This process has gone furthest in countries such as Germany, the Netherlands and Britain whereas retailing is still predominantly a preserve of small independent retailers in countries such as Japan, Spain and Italy. In the UK, clothing retailing is more concentrated than in the rest of the EC, and it is rather more concentrated than the UK retail sector as a whole. Firms with 100 or more outlets account for 48% of turnover of clothing products (compared with 47% for all retailing).[8] Marks and Spencer and the Burton Group alone account for about 25% of clothing sales. Insofar as large retailers buy from British suppliers, this has helped the UK industry to survive. But this has been at the cost of long-term adverse effects, most notably in shifting many of the fundamental decisions concerning product development and product design into the retail organizations. This is particularly evident in clothing manufacture where, in effect, many clothing manufacturers have become sub-contractors, and rely heavily on orders from large retailers. Where retailers shift to greater use of overseas sourcing, manufacturers can find that they are ill-equipped to compete in other markets.

Concentration in retailing has been paralleled by an element of concentration in the upstream parts of the chain, most evidently in the rise of large transnational companies which have subsidiaries operating in all stages of the production chain.[9] However, for a number of reasons, increased concentration has been followed by an increase rather than by a reduction in levels of competition. In part, this is a matter of market maturity; the percentage of household income allocated to clothing (and footwear) has fallen from an average of 11.8% in 1953/4 to 7% in the late 1980s.[10] This shift in spending patterns offsets much of the effect of rising incomes so that, in many cases, firms can only increase sales at their competitors' expense. Despite the growth of the large chains, there remains a significant small business sector. In 1988, some 22% of turnover in clothing was taken by retailers with only one outlet.[11] Clothes and other textiles are often marketed on street stalls while low capital costs for starting small retailing units provide a continuous potential for new businesses, some of which may aim at specialized areas of the market. At the other end of the scale, large retailers whose main activities lie in other sectors, for instance, Tesco, Sainsbury and Asda, have shown a capacity to invade the market for textile products – although primarily in the lower priced areas of the market. Additionally, the sector is an attractive one for mail order firms which had some 10% of total UK clothing sales in the 1980s.

The intensity of competition is also sustained by diversity in the markets for textiles. This limits the advantages that can be gained from large-scale production. This is particularly so in clothing products where consumer pursuit of individuality offers firms a seemingly endless potential to extend the levels of market segmentation and to generate product redundancy. Strategies aimed at segmentation emphasize the extension of product variety, and increasing product differentiation, particularly through greater diversity in styling and in quality. In both cases, an important role has been played by the development of a variety of new fibres and fabrics which have contributed to the emergence of new ranges of specialized products, for instance, in the sports and leisure sector.

Where product redundancy is concerned, the role of fashion is pivotal. Manufacturers and retailers have sought to extend a fashion element, and thus an implicit redundancy, to a wider range of products. This applies to basic household textile as well as to clothing and soft furnishings. The consequences of this in terms of the retail lives of products is indicated by a survey of the American market, summarized in Figure 5.[12] Some 35% of products in the US market were identified as fashion items, and a further 45% as seasonal so that about 80% of the market is for products with a shop life of twenty weeks or less. The pattern in the UK appears to be similar. In part, this has been accomplished by an increasing detachment of 'seasons' in the retail sense from the more traditional seasonal delineations. An increasing number of retailers have anywhere between five and ten 'seasons' in a year.

Thus, patterns of market change seemingly emphasize rapid rates of

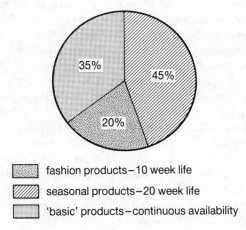

fashion products–10 week life

seasonal products–20 week life

'basic' products–continuous availability

Figure 5   Apparel product life, USA.

product change and increased product diversity. In manufacturing terms, this indicates a shift towards production of wider product ranges in lower volumes – using the flexible production strategies considered in Article 3.7. However, it is possible that the market for clothing is in the early stages of a major change. With some notable exceptions – such as Benetton and C & A – retail firms have tended to operate within national frontiers, but this is changing as large retailers seek to extend their operations across the European Community and, in some cases, across the markets of the industrialized countries as a whole. To the extent that international retail firms are able to sell similar products in different national markets, the potential for long production runs may increase. It also seems probable that spending patterns on clothing, as for other products, may be modified by increasing concern with the global environment. This is likely to militate against fashion-driven consumption patterns. Further, levels of economic growth seem likely to be lower than in the previous forty years, emphasizing the virtues of garments designed on 'classic' lines for long lives.

■ Changing patterns of supply

Competition has also been intensified by growing diversity in sources of supply. This is associated with the growth of manufacturing capacity in the developing countries (DCs), and has contributed to the expansion of global capacity to levels which exceed product demand. In several respects, textiles provide an ideal target for an industrializing country, particularly in the early stages of industrialization. The capital requirements for new firms are limited in some parts of the sector and, because of the fundamental nature of textile products for most human societies, there is generally a reservoir of many of the skills required to develop industrial systems for spinning, weaving and apparel making. The varied nature of the industries within the textile sector (see previous section) provides opportunities for subsequent upstream development in the more technologically and capital intensive parts of the sector, such as fibre production and equipment manufacture.

The growth of DC production and entry into the international market has been assisted by labour costs which are often very far below those in the industrialized countries (ICs). For instance, a survey by Werner International in 1987 found a variation in labour costs in spinning and weaving ranging from $15.70 per hour in Switzerland and $12.98 in (West) Germany to $0.23 in China and $0.20 in Indonesia. Low labour costs were generally accompanied by longer working hours, ranging from an annual average of 2092 hours in Switzerland and 1811 in Germany to 2464 hours annually in China and 2447 in Indonesia. Britain was towards the mid rank position (11th) in terms of hourly labour costs ($7.09), but annual working hours of 1775 were below the average.[13]

Rising DC production has been met by a range of responses in the industrialized countries. A central element has been trade protection. First introduced in the early 1960s as a 'short-term' measure to assist adjustment within the industrialized countries to changing trade conditions, this temporary expedient is set to last well into the twenty-first century in its current guise of the Multi-Fibre Agreement (MFA). Paradoxically, the MFA was established through the General Agreement on Tariffs and Trade (GATT) which was established to liberalize world trade. The MFA rules have become increasingly complex and comprehensive in their potential coverage, and they enable ICs to establish quotas and other restrictions where it is believed that domestic producers are being — or might be — 'seriously damaged'. However, the MFA's effect has been to retard rather than inhibit the growth of DC exports.

A second strategy for IC firms has been to relocate parts of textile production to low labour-cost countries by either establishing production facilities in favoured DC sites or through sub-contracting and buying arrangements. By the 1980s, the global production and buying network encompassed not only relocation from the ICs but relocation from newly industrialized countries such as South Korea and Taiwan, for whom growth has brought its own problems of rising labour costs and impaired competitiveness. Much of the global network is located in the near 'periphery' of areas adjacent to the major industrialized areas, notably the Mediterranean, the Caribbean and South East Asia. Tax regimes in countries such as Germany and the USA have allowed labour-intensive elements of production to be undertaken 'off-shore' while encouraging retention of activities which generate high levels of added value — essentially those which are knowledge and skill intensive. For instance, many German apparel manufacturers have retained apparel design, cutting and finishing in domestic locations while sending garment components for the labour-intensive assembly processes to factories in Eastern Europe. The proximity of countries such as Turkey, Romania and Bulgaria keeps transport costs down and avoids the time penalties and other problems which can be associated with sub-contracting to firms in more distant locations.

A third element of strategy comes from 'capital deepening' – particularly through investment in the development and application of labour-saving technologies. As is discussed in Article 3.7, this has proved to be very difficult in some areas of production. But where automation or semi-automation has been achievable, the cost structure of production can be transformed, through the reduction of the proportion of direct labour costs and other savings. For instance, the introduction of computerized systems of pattern grading and marker making in clothing manufacture was found to reduce labour costs in these activities by about 40%, and to lead to improvements of up to 9% in levels of fabric utilization (fabric accounts for about 50% of costs in clothing manufacture).[14] International surveys of manufacturing costs have shown that, in some cases, capital intensive production methods can reduce labour costs per unit of output and total manufacturing costs in ICs to levels below those in developing countries. For instance, unit labour costs in US spinning and weaving were found to compare favourably with those of India and Brazil; high productivity had offset high labour costs.[15] But, until flexible, programmable equipment became available (see Article 3.7), this strategy could generally be applied only to high volume production of standardized designs.

At one time or another, strategies by governments and by companies in all the ICs have placed emphasis on automation and on the concentration of production in large units in order to achieve the maximum economies from large-scale production. But the associated cost structure, and other factors such as de-skilling and reliance on high-cost equipment dedicated to single or a limited range of tasks tends to result in inflexibilities and inability to respond to changing patterns of demand. It has also locked companies into standardized product lines where margins are low, and where DC firms are most able to compete, particularly where they utilize similar types of automated equipment. For instance, automation in some areas of cotton spinning has so de-skilled the process that labour costs, even though now a low proportion of total costs, have remained a primary determinant of location.

The problems associated with such 'Fordist' production methods are not confined to the textile sector and, as in other sectors, there has been a search for alternative strategies. In a number of countries, notably Japan, Germany and Italy, governments and firms have more readily identified the problems of over-concentration and developed new strategies. In part, these have centred on shifts to product specialization, partly through the devolution of production to more adaptable and flexible small and medium-sized units and firms. The resources available in traditional textile communities have also been important, providing a foundation for upgrading workforce skills and experience. It was realized that these skills, rather than the use of automated equipment which is usually rapidly available across the world, offers a potent source of distinctive competence. Concentration on the more skill intensive, higher value areas of the market has paid off in trade terms, as in other respects (particularly those of human worth – see Article 3.7). Thus, the German deficit on textile trade was contained while Japan remained in surplus. Italy provides the most striking case in this respect. By the mid 1980s, the textile and clothing industries accounted for 10% of Italian gross domestic period, and 21% of Italian exports. Italy became the world's largest net exporter of clothing, with a trade surplus of $6.4 billion on this item alone, and with a total textile and clothing surplus of $9.9 billion – achieved, it should be added, with labour costs only slightly below those of West Germany, and with shorter annual working hours.[16]

Thus, rising textile imports in the ICs are neither inevitably nor solely a consequence of Third World industrialization. Specialization within the industrialized countries has led to increased trading between them. For instance, Western Europe accounts for about half of the world's textile trade, but 80% of this trade takes place within Western Europe.[17] In the case of the UK, imports of textile products rose from about 12% of domestic consumption in the late 1960s to more than 30% for clothing and over 40% for other textiles by the 1980s when the deficit on textile trade reached $4 billion. But the proportion of imports (by value) which came from other West European countries was similar to that from the DCs. What is more important than the level of imports is the extent to which imports are offset by exports which are underpinned by long-term product and manufacturing strategies geared to areas of maximum advantage. It was shortcomings in this respect rather than competition from the DCs which constituted the major UK problem for the textile sector.

How the pattern will alter in the future depends on a number of factors, including the outcome of negotiations in the early 1990s to replace or, possibly, to phase out the MFA. What will also be important is the extent of a shift within the industrialized countries to changes in the overall supply system for textiles towards a 'quick response', demand-driven system. This will require some fundamental changes in the organization and functioning of the overall textile supply chain. (These issues are considered in Article 3.7.)

Notes

(1) Office of Technology Assessment (OTA) (1987) *The US Textile and Apparel Industry: A Revolution in Progress.* Washington DC: US Department of Commerce.

(2) This is a simplification. Mechanization in the textile industry was a primary factor in the growth of the chemical industry in the late 18th century, for instance in the rise of production of sulphuric acid, alkalis and other processing materials, and then in the late 19th and early 20th century growth of 'synthetic' dyestuff manufacture (as opposed to the use of naturally occurring dyes from plant and other sources). The past 50 years have seen an intensification of this interdependence, particularly through the

manufacture of synthetic fibres, and improvements in the processing of textile materials (see Article 3.7).

(3) The only other sector where retailing has a comparable role is food manufacture.

(4) Central Statistical Office (1980) *Annual Abstract of Statistics*. London: HMSO; also 1991 edition.

(5) Clairmonte, F. and Kavanagh, J. (1981) *The World in their Web*. Zed Press.

(6) Figures from Ministry of Labour Gazette and Employment Gazette. They are approximate figures because there have been changes in the groups included in these categories.

(7) Central Statistical Office (1988) *Business Monitor: PAS 4555, Retailing*. London: HMSO.

(8) Central Statistical Office *Business Monitor, op. cit.*

(9) However, compared with most other UK manufacturing sectors, textiles has a relatively low level of concentration, particularly downstream from fibre production, and most notably in clothing manufacture where there are many small firms.

(10) Central Statistical Office (1989) *Family Expenditure Survey 1988*. London: HMSO.

(11) *Business Monitor, op. cit.* The figure for clothing is rather lower than the share of single outlet firms of retail turnover as a whole – 28%. On the other hand, single retailers accounted for larger than average shares in household textiles and soft furnishings (32.2%) and in carpets, carpeting and rugs (43.3%).

(12) OTA (1987) *op. cit.*

(13) Quoted in Anson, R. and Simpson, P. (1988) *World Textile Trade and Production Trends*. Economist Intelligence Unit, Special Report No. 1108. Germany was 6th in the rank order by wage levels among the 30 countries surveyed.

(14) Hoffman, K. and Rush, H. (1985) *Microelectronics and Clothing: The Impact of Technological Change on a Global Industry*. Science Policy Research Unit, University of Sussex.

(15) Anson and Simpson, *op. cit.*

(16) Anson and Simpson, *op. cit.* The Werner estimate of labour costs may be on the high side because of the problems of measurement in Italy's extensive 'informal' sector, an important factor in its success in this and other industries.

(17) Anson and Simpson, *op. cit.*

2.9

Design Project Planning: Case Study of a Litter Bin

T392 course team

[This takes another view of designing, this time emphasizing the planning that needs to be done. The approach to planning the process is illustrated in the context of a particular company intent on tendering for the contract to make litter bins.]

This example of an industrial project is concerned only with the planning and execution of the design phase of a production contract (i.e. all the preparation for production, but not the production itself). There is much in common with design activities for all kinds of products so there are general lessons to be learned from this example. By contrast, production plans are nearly always highly specific to the product in hand, so are not dealt with here. Nor is this example concerned with the detail of a particular design; it is, rather, concerned with illustrating how the planning of a design project might progress.

The company in question was well established in the fields of general engineering and sheet-metal working. Its main strength was in the production of various storage and processing tanks with capacities ranging from a few litres to many gallons. It had a wide experience of specialized finishes, including galvanized and tinned steel, epoxy and rubber linings and paint and plastic coatings. It had also experimented with the production of vacuum-formed plastic and glass-fibre reinforced polyesters. The opportunity arose for the company to tender for the supply to a local authority of a general-purpose litter bin. The project looked attractive because it could be a profitable continuing line within the company's present expertise (much to be preferred to diversifying into something completely new in the existing economic climate).

■ Getting started

For their own designs the company usually called upon the services of a consultant design engineer, who could be relied on to deliver well-engineered designs suited to their methods of working. When asked to take on this project, modest though it was, the designer thought that it presented a number of new problems for the firm (such as competing in an unfamiliar market and meeting new aesthetic standards) and that it required a rather rapid response. He also accepted advice given to him that design by a team rather than by individual specialists was likely to be more successful in winning markets, so he persuaded the company to allow him to form a design team to work with him.

However, it was felt that the composition of the design team should be appropriate for the project, and this demanded some further information. The following is part of the list of questions he drew up which are not only of interest to him as a designer, but which also have a bearing on the kind of team that he should form.

- What is the patent situation?
- Are there export possibilities?
- Are there regulations as regards safety, fire hazards, etc.?
- Are these regulations the same in other countries?
- Are there other regulations about 'street furniture'?
- What about appearances? Is there a preference for a particular style, or are several alternatives desirable?
- Is transport of the units an important consideration?
- How are the units to be emptied?
- Is there a maximum price that the council is likely to accept?
- What production methods are acceptable for the company, and is capital investment in new equipment going to be needed?
- What should the bin capacity be?
- Should the bin be fixed or moveable? If moveable, who is responsible if the bin falls into the road?
- How can the problems of vandalism, and the consequent danger to the public, be reduced?
- Is capital likely to be available on special terms for a contract with social benefits like this one?

It was obviously important for the design team to include people who could handle the answers to all these questions, or produce well-informed answers to them if, as was likely, the local authority was prepared to consider sensible suggestions for any of them. And yet it was also important to keep the team small.

■ Forming the design team

The designer regarded the composition of his design team as crucial to the success of the project and so he devoted a good deal of thought to it, and then made sure that he got his way. To help organize his thoughts on the matter he used the simple technique of drawing 'bubble diagrams'. Figure 1 shows the 'bubble diagram' he drew while trying to plan the kind of team he needed. A bubble diagram is simply a convenient way of organizing one's thoughts about how a complex topic or project should be divided up in order to be dealt with in smaller chunks – in this case by several people. Here, each bubble represents an area of concern that needs to have work done on it.

To deal with these four sub-areas of concern within the overall project he invited the following people to join the design team, which he would chair:

(1) A junior consultant colleague – to investigate any constraints as regards safety of pedestrians and traffic, fire (especially of litter in the bin), patents, etc.; and to visit two other countries, one of which should be

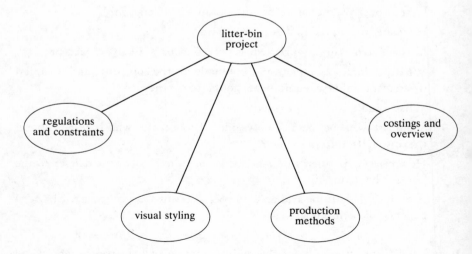

Figure 1   A bubble diagram to identify key aspects of the litter-bin project.

Switzerland because of its reputation for tidiness, to find out their regulations and photograph some of their designs.

(2) A graduate from the Royal College of Art who had specialized in relevant aspects of industrial design — to attend to the aesthetic aspects of the design; to help produce mock-ups; to join with the designer and the marketing manager in preparing the presentation to the local authority.

(3) The production manager of the company — to advise on production methods and materials and costs and the possibility of new methods being introduced into the company; to deal with prototype construction.

(4) In the absence of a marketing manager in the company, the managing director — to take an overview; to comment on the possibility of capital investment in the project with an eye on possible future projects; to keep an eye on the dangers of overspending.

The designer felt that his own experience would enable him to keep good control of the project as a whole and enable him to integrate the various inputs to the team and create a successful design.

He realized that as funds were limited the two external consultants would have to be offered contracts which depended partly on the success of the tender. That is, they were offered (and they accepted) a fairly low flat rate plus a substantial bonus if the tender was successful. Evidently, if a really good design could attract an extra £1 per bin, or could save £1 per bin in production costs, or both, the consultants' fees would easily be covered. Both consultants were invited to help with the eventual presentation of the chosen design to the local authority.

■ Planning the agenda and the timing

The project set out with a clear time limit of three months, within which it was necessary to create good designs and prototype models. Obviously the limited timing called for a tight schedule and a clear agenda of work. Again it was necessary to break the project down into sub-projects, but this time each sub-project was to refer to a stage of the project rather than a specialist field. It was expected that each member of the team would have a contribution to make to each stage. Thus, whereas the first bubble diagram could be thought of as 'vertical' or specialist slices through the project, what was now needed were 'horizontal' slices through the project distinguishing between successive stages of progress. Figure 2 shows the bubble diagram that resulted. The ordering of the 'bubbles' indicates the order in which the different aspects of the project should be tackled.

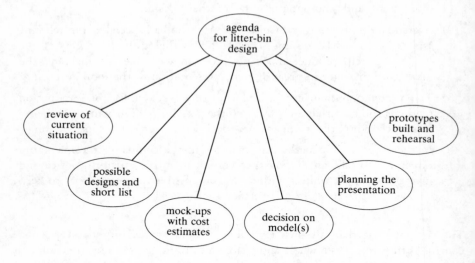

Figure 2   A bubble diagram to generate agenda headings for the team meetings of the litter-bin design team.

The designer arrived at his subdivision of the project simply by using his experience to judge how he thought events could most efficiently proceed. He felt that the presentation was a vital step in winning the order and wanted to create enough space in the timetable to be well prepared when the time came, not only with attractive prototypes but also with well-informed answers to all the likely questions that might be asked.

Within each sub-project the designer identified the following tasks and linked them to those on the team who would be primarily responsible for them:

(1) *Review*   Summary of all the regulations, patents, codes of practice both in Britain and in some of the possible export markets (consultant). Information on existing designs and public comments, if any, on them (RCA graduate). Further guidelines, if any, from local authorities who have awarded similar contracts, e.g. preferred emptying methods, etc. (chair). Financial statement (managing director). This phase to conclude with a specification of the required (and desirable) characteristics of the litter bin.

(2) *Possible designs*   A brainstorming session with prepared inputs from each member of the team. Two meetings, or two parts of one meeting, would be needed: one to be creative, the other to arrive at a short list of designs.

(3) *Mock-ups and cost estimates*   Short-list to be worked on, and further reduced for technical, production, regulatory, aesthetic, or cost reasons, etc.

(4) *Decision meeting* Choice of the final design, or possibly more than one design (e.g. one for a covered shopping precinct and one for out-of-doors). If two designs are chosen, they should preferably have a good deal in common from a production point of view.

(5) *Planning the presentation* Who should say what, what displays should be made, how the prototypes should be displayed, timing, etc.

(6) *Rehearsal of the presentation* (Other members of the company could perhaps represent the local authority.)

The timing of all these events is represented in the simple bar chart of Figure 3.

At the first meeting scheduled in this chart, the chart itself and all the stages of the project, as set out above, would be discussed, modified if necessary, and agreed. Each member of the team would then be able to draw up a clear workplan covering the next three months.

In practice, this workplan was, broadly speaking, accepted, except that two additional 'working groups' were agreed to. First, the designer from the RCA thought she would find it very helpful to meet more frequently with the chair, in order to ensure that her designs were consistent with the existing working method within the company. Secondly, the production manager felt that he and the chair should work in more detail on the designs of the various parts of the bin (e.g. the lid, if any, the inner and outer containers, the method of fixing where required, materials, finishes, safety factors, etc.). This completed the planning of the project.

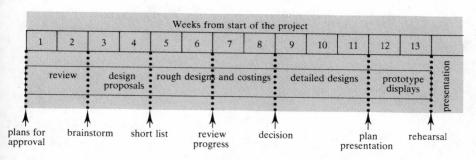

Figure 3 A single line bar chart displaying the plan of action for the litter-bin project.

PART 3

Specific Technologies

Introduction to Part 3

R. McCormick, C. Newey and J. Sparkes

As we noted in the introduction to the whole Reader we have tried to select technologies that have some connection to school technology, while maintaining the 'real-world' flavour. We have only been able to include a small selection of specific technologies in this section of the Reader. The selection tries to sample the exotic, such as the compact disc (exotic but nevertheless familiar to many pupils), the simple, such as a cardboard skeleton, the familiar, such as bread (although the industrial production of it is unlikely to be familiar), and the complex, such as air traffic control systems. None of these are included because they can be translated into school technology, but rather as a means to explore the nature of technology in action (inasmuch as written accounts can). It is this understanding of the nature of technology that will form the basis for selections of ideas and materials for inclusion in schools.

3.1

The Fisher–Miller Skeleton

D. Walker

[This article is an extract from a case study of an idea for the development of a skeleton for use in schools, which came to fruition and became a commercial enterprise. The two people involved in this project are Richard Miller, a primary school teacher, and Bryan Fisher, the ex-Managing Director of a company that manufactured plastic toys. We enter the story of the project at the point where Richard Miller had identified a need within his class situation.]

■ The designer input

During his work as a teacher, Richard Miller had wished to explain human anatomy to primary school children. He discovered that he could borrow a skeleton from his local education authority (LEA) but there was a six month waiting list. He began to think of constructing a skeleton with methods that could be used by his own class. He started to look systematically at the appropriate techniques and was particularly influenced by packaging, such as McDonald's Apple Pie and Toblerone. He saw that the triangular form of this latter packaging approximated to the cross-section of a leg bone. He bought a second-hand copy of *Gray's Anatomy*, borrowed a half skeleton and began to construct a skeleton piece by piece.

Like many designers Richard Miller was in the position where he had to devise the process as well as imagine the finished article. He started by making copies of human bones at life size. He began to experiment, first with papier mâché then with plasticine (which he kept in the fridge in order to make it easier to shape by carving), and then by paper folding. He realized early on that an exact copy could only be achieved by some kind of casting process, but that an approximate shape could be arrived at by folding card. A friend of his owned a monkey skull which has a pronounced central ridge running down the cranium, and actually looks as if it is formed by the folding of two planes. Triggered by these perceptions and with the examples of certain kinds of packaging before him, Richard shaped the frozen plasticine bones with a flat knife which gave them a series of facets. This was a crucial step. Although an element of approximation was involved, he now had bones that he could convert to a series of flat shapes. In order to discover the exact shape of each faceted plane, he wrapped the bones in tissue paper and marked all edges and creases, and then unwrapped the tissue and laid it flat. The basic patterns of the bones lay in front of him.

This technique, probably unique to Richard Miller, can be quickly described in a few words, but this glosses over the hours of trial and error spent in shaping the bones, experimenting with the tissue, making prototypes and making the detail accurate enough – in short, making the technique work.

In the early days he was motivated by curiosity and the challenge of solving the problem. But as he worked two other influences came into play.

> 'It was always at the back of my mind since art school. There I saw lots of students trying to do one-off fine art pieces of sculpture and struggling to get them seen by anyone – and they would go off and make adventure playground items or dinosaurs for theme parks, or whatever. I had the feeling that it was no good living in a garret making one-offs and hoping to be discovered. One had to do something else. And my idea, if you like, was sculpture which could be packaged flat and assembled by a customer.'

Another influence was prompted by an article in *The Guardian* at the time he was working on his prototypes. It reported that skeletons, normally from the Third World, were now difficult to obtain. Correspondingly, the cost of skeletons in the West had risen to £300. Even plastic facsimiles cost about half that figure. Slowly the idea grew that the cardboard skeleton might have a big market potential. At this stage he thought that there might be an audience beyond schoolchildren, perhaps in the medical field.

These thoughts were rather unfocused, because the challenge of the problem and self-education were paramount in Richard Miller's mind.

> 'I would say business considerations were very secondary to me. It was more a compensation for not having been able to do biology at school and it rounded out my knowledge of anatomy.'

■ The business input

Richard Miller's work may not have moved much beyond the prototyping stage had it not been for his contact with Bryan Fisher. Although his first business experience was in plastic manufacture and engineering, Bryan had come to understand very well the toy and educational games market. He also knew the stimulus to business that could come from a radically new, well thought-out design. He was predisposed to treat design seriously and was able to judge its potential. He was invited to visit Richard Miller and found in his living room a full-scale cardboard model of an adult human skeleton. Having seen this, Bryan Fisher said, 'I was stunned. I was completely bowled over. I thought it was outstanding. I thought "it has got to be made".'

After the first meeting between Richard Miller and Bryan Fisher, they decided very quickly to go into partnership. Bryan Fisher's impact on the project was dramatic. Whereas Richard had conceived the skeleton as a handicraft product for schoolchildren, with italic script captions and shapes to be cut out with scissors, Bryan saw it in more ambitious terms. From his previous experience he knew what was required to achieve commercial success on the basis of an original design.

He saw immediately that the skeleton potentially had a domestic and educational market *world-wide*. He also knew that the handicraft idiom was to be avoided. The skeleton, he thought, should be as easy as possible to assemble. His impulse was to go for mass production. This meant three things for the designer, all of them entailing further work:

- re-labelling all the parts with transfer lettering
- providing complete sets of instructional drawings
- making drawings for a creaser-cutter process (an industrial process common to folded and card products).

Bryan Fisher knew that the instructional drawings were crucial to overseas success. He wished to avoid the complexities of translating verbal instructions into a variety of languages, and drawings are an international language. It was also crucial, in Bryan's view, that they invest in a creaser-cutter machine which would make mass production possible, give a very accurate model and remove much of the labour in assembly. Under Bryan Fisher's prompting, Richard Miller flew to New York to see a business consultant. He endorsed the three points above and made a further helpful suggestion – that, as a way of testing the market, they should launch the skull on its own. Clearly this meant much less initial capital investment, less risk and less effort.

■ Product launch

The skull was launched in 1980. From his previous experience Bryan Fisher had discovered that mailing shots had a poor response and that single advertisements were 'a waste of money'. Accordingly he set out to attract press attention to the skull, and achieved this aim in *Design* magazine in 1980 and *The Observer* colour supplement, 28 June 1981. As a direct outcome of this publicity, the first run of 2500 skulls and a second run of 5000 were quickly sold out. Later,

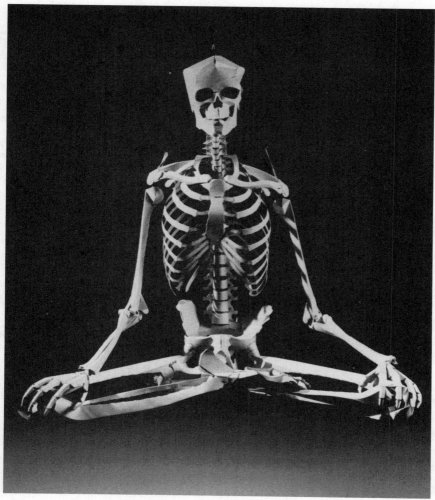

Figure 1 The Fisher–Miller skeleton.

on the launch of the full skeleton, Bryan Fisher stimulated even greater press coverage. He gave away full skeleton kits to four national newspapers. The skeleton attracted a full double-page spread in *The Observer* colour supplement. For the cost of the first four give-away skeletons and two further skeletons for photography (roughly £210), Bryan Fisher achieved coverage which, as an advertisement at that time, would have cost around £20 000.

[We leave the case study there. The skeleton provided just a starting point for a whole business enterprise. The full case study can be found in OU Pack P791 Managing Design.]

3.2

Food Production: Bread

M. Palin

■ Introduction

Cereals have been cultivated as sources of food for around 10 000 years. They are eaten as dense foods such as porridge or boiled rice, or more elaborately processed into baked products such as bread which are much less dense.

Only wheat and rye can be used to make the less dense products because of the physical and chemical properties of their proteins. Wheat produces what is generally perceived to be the highest quality products and consequently wheat is produced in the largest quantity.

Flour products and in particular bread have been staple foods in the United Kingdom for centuries, although consumption has been declining for many years. Lower income households consume, on average, significantly more flour products than higher income families.

The proportion of white to brown and/or wholemeal bread eaten has varied considerably apparently depending on factors such as price, fashion and nutritional advice. Current dietary recommendations encourage higher consumption of cereals, particularly whole grain cereals and this is probably a major factor in the current rise in the amount of wholemeal products being consumed.

In the United Kingdom the majority of bread is produced by large-scale plant bakeries, a small proportion by small bakers and an increasing amount by in-store bakeries which are a significant growth area. The retailing of bread is also provided for in a variety of ways with the supermarkets accounting for a large proportion of both the plant bakeries and of course the in-store bakery output. Specialist bread and confectionery shops, many of which are owned by plant bakeries, are also significant retailers.

■ Wheat grain

The wheat grain is an ovoid body measuring 5–8 mm in length and 2.5–4 mm in width. It has a deep crease along one of the long sides. The grain surface is covered by several layers of protective brown tissue known collectively as bran which accounts for 14–16% of the weight. At one end of the grain is the germ (2.5–3.5%). Under the bran is the white endosperm which forms the bulk of the grain (81–84%).

In order to produce wholemeal flour the grain must be crushed into a powder. For white flour most of the bran must be removed before crushing, and the germ is also removed in the process. Brown flour is made by adding back some of the removed bran.

■ Types of wheat

Wheat can be classified as being hard, soft, strong or weak. Hard and soft refer to the way in which the grain reacts to milling; in hard wheats the endosperm tends to break into more regular-sized pieces and to separate more readily from the bran. Strong and weak refer to the breadmaking characteristics of the wheat. Strong wheats generally have a higher protein content and make bread of large volume, good texture and good keeping quality.

In general, hard wheats produce strong flour suitable for breadmaking, whereas soft wheats produce weak flour more suitable for cakes and biscuits. Wheat grown in the United Kingdom tends to be softer than wheat grown in hotter, drier climates such as Canada, and traditionally breadmaking flours used in the United Kingdom have contained a significant proportion of imported wheat. Modern breadmaking practice has reduced the need for imported wheat although it is still widely used by smaller bakers and by those producing premium quality bread.

■ Milling

The milling system for making white flour may be described as a series of interlinked grinding and sieving operations aimed at extracting the maximum amount of white flour that is free from contamination with bran specks. The system aims to open out the grain often along the crease, then to detach the endosperm from the bran and gradually to reduce the particle size of the endosperm.

The steel rollers used are of two different types: the break system and the reduction system. The break system consists of a series of pairs of fluted rollers. The grain passes between the pairs of rollers which separate the endosperm from the bran. The reduction system is a series of pairs of smooth rollers which crush the endosperm. In both systems the gap between the pairs of rollers is progressively reduced and repeated sieving occurs. This helps to separate the bran from the endosperm and to reduce the energy expended and to increase throughput.

□   **Extraction rate**

The ratio of the weight of flour obtained to the weight of wheat milled, expressed as a percentage, is called the extraction rate. Although about 82% of the wheat grain is endosperm, the maximum extraction rate for white flour is around 75%, brown flour is generally around 85% extraction, and by law wholemeal must be 100% extraction. The actual extraction rate achievable for white flour depends upon the quality of the wheat and upon milling factors such as the degree of wear on the rollers.

□   **The breadmaking properties of wheat flour**

When wheat flour is mixed with water to form a dough, the protein absorbs water and forms the cohesive, elastic and extensible substance called gluten. The breadmaking quality of any flour is a reflection of the quantity and quality of the gluten it can form. In general this depends on the quantity and quality of the protein in the flour, i.e. strong flour is higher in these properties than weak flour.

Another important consideration is the amount of enzyme activity, particularly alpha amylase, which breaks down starch. If this is too high the crumb becomes sticky and may collapse and the crust browns too much when baked.

The higher the water-absorbing capacity of a flour, the greater the quantity of dough and therefore of bread that can be produced from a given quantity of the flour. The amount of water absorbed during the formation of a dough of satisfactory consistency depends on the flour's protein content, its enzyme activity and the degree of mechanical damage sustained by its starch granules during milling. The more damage, the higher the absorption of water. The starch of hard wheats sustains more damage than the starch of soft wheats, and the amount of damage can be increased by adjustments to the milling process.

■   Breadmaking

Bread is made from a basic mixture of water, flour, yeast and salt. Various other substances are used from time to time, e.g. vitamin C (abscorbic acid) is used in most bread produced in a plant bakery. Enzymes from the yeast can convert sugar in the dough into carbon dioxide gas and ethanol. The carbon dioxide causes the elastic dough to rise. Most of the ethanol evaporates during baking.

☐ Traditional breadmaking

Many differing processes can be used for making bread. The process most commonly used in the United Kingdom until the early 1960s was the bulk (long) fermentation process. The procedure outlined below was (and is) used for most traditionally produced bread in the United Kingdom but many variations are used, e.g. time and temperatures can be altered.

In the bulk fermentation process the ingredients are mixed in a low-speed mixer for 15 minutes and the dough is then set aside for three hours at a temperature of 27°C. A longer fermentation can be produced by lowering the temperature. After two hours the dough is briefly mixed again to bring about better contact between yeast cells and soluble sugars in the dough. After the third hour of fermentation the making up of the dough begins, in preparation for baking. The dough is divided into pieces of the desired weight, each of which is moulded into a ball. After 15 minutes resting, the piece is remoulded and placed in a baking tin. Fermentation continues for 45 to 50 minutes at 38–43°C in a humid atmosphere to keep the dough surface moist and so prevent a skin forming which would inhibit rising. The dough rises as the carbon dioxide develops.

The dough is then baked in an oven at 235°C for 25 to 40 minutes. As the temperature of the dough rises the fermentation rate increases and the gas bubbles expand. Eventually the temperature of the dough rises sufficiently to denature the enzymes and fermentation ceases, the starch granules swell and the gluten network coagulates. This holds the gas bubbles in their final positions and sets the dough in its final size and shape.

After removal from its tin the loaf is cooled for two-and-a-half to three hours before slicing and wrapping. Wrapping when still warm means that moisture evaporating from the cooling bread condenses inside the package making the crust damp and tough.

Several variations of this basic process exist, often in particular regions or countries. A major variation is to mix the yeast and some water with some of the flour and allowing fermentation to take place before combining this with the rest of the ingredients.

☐ The Chorleywood bread process (CBP)

This process was introduced in the early 1960s and is currently used to produce the large majority of bread produced in the United Kingdom. The keypoint of the process, which distinguishes it from the traditional method, is the replacement of the long bulk fermentation by a short period of high-speed mixing. So the changes which occur in the first two hours of fermentation are replaced by mechanical energy generated by machinery and this is used to modify the structure of the dough. It is a semi-continuous process in that the dough is mixed in batches but from then on the process is continuous.

The amount of mechanical energy put in is critical. 40 kj/kg dough must be put in within two to four minutes using a special high-speed mixer. Compared with the traditional mixing process, five times as much energy is used in one-fifth of the time. However, because a large part of the energy use is during baking the CBP only increases energy use by about six per cent. After mixing, the dough is treated in the same way as after the third hour of fermentation in the traditional process.

Various recipe modifications are required if satisfactory bread is to be obtained. These include the addition of vitamin C (ascorbic acid) and the use of more yeast and water. However, there are considerable advantages to be gained for a bakery in adopting this method:

(1) 60% saving in total processing time;

(2) 75% saving in the space needed;

(3) 75% reduction in the amount of dough in process at any time so that in the case of a breakdown the losses are less;

(4) 4% increase in yield;

(5) flour of about 1% lower protein can be used.

■ Quality maintenance

Manufacturing industry has traditionally used quality control procedures to ensure that sub-standard goods do not leave the factory.

Most companies today use the rather broader concept of quality maintenance. This centres around the notion that fault prevention is better than fault rejection. It uses a systems approach which views the whole process as a single entity. This allows information to be passed backwards and forwards through the system so that quality can be maintained more efficiently. The breadmaking system is no exception to this.

□ Quality maintenance of flour

The miller's critical quality control procedures centre around the various wheat supplies purchased and includes detection of contamination, e.g. by stones and weed seeds as well as measurement of protein and enzyme activity. A portion may be test-milled to evaluate its milling quality, achievable extraction rate and baking quality to enable an appropriate grist to be decided.

During the milling process, checks are kept on the output at various stages. In this way it is possible to have an early indication of the requirement to adjust machinery. It also provides the necessary information for the miller to produce different grades of flour by blending the output from different parts of the mill and from different grists.

The factors affecting the breadmaking properties of flour were discussed earlier. It is necessary for the miller to monitor these factors so that a flour of consistent breadmaking quality is supplied. Tests are available to measure protein content, water absorption content and alpha amylase activity. Protein quality is best assessed by making bread using a standard procedure although procedures can be applied to a dough which gives some indication of protein quality.

There are regulations requiring various vitamins and minerals to be present at specific concentrations in most flours. The flour producer must monitor this and make the appropriate additions.

☐   **Quality maintenance of the breadmaking process**

Most of the following section would apply to any manufactured food as in most foodstuffs quality control involves both subjective and objective measurements. Probably the most important objective measurement is that of weight.

Most bread is produced to a legal minimum weight. The economics of producing a high volume, low profit-margin product like this means that machinery must be set to minimize the 'overweight' on each item. This reinforces the need for careful monitoring as the margin of error is slight.

Although it is the selling weight that is the critical one, the main monitoring point is seen after the dough has been divided. Appropriate dough weights are calculated taking into account the weight loss that will occur during the rest of the process. The advantage of monitoring the weight at this point is that it allows the dough to be recovered and redivided to the correct weight, thus eliminating waste. Check-weighing after baking is often carried out as a final precaution against underweight product being sold but, if this was used as the main point of control, any fault in weight would take a long time to show up and a considerable amount of underweight bread might have to be thrown away.

In modern plant bakeries, weight control is highly automated and under computer control so as to provide a constant record of product weights and to immediately reject any underweight dough pieces. The control system will also indicate overweight dough pieces and alert staff to the fact that adjustment is needed. 'Giving away' bread reduces profit on a product which in any case carries a low profit margin.

Metal detection apparatus is used so that an individual loaf registering the presence of metal is rejected automatically. Other foreign bodies are much more difficult to detect and so precautions to avoid contamination must be carefully devised and strictly observed. Other objective testing can include measurement of loaf volume and of crust colour.

Subjective testing would normally be carried out on random samples. As such testing is often destructive and thus renders the product unsaleable, only a relatively small number of items are chosen. Although this means that the procedure is not statistically valid, it is usually regarded as important that such things as taste, smell and appearance as well as physical properties such as ease of slicing and softness are examined regularly. In the same way as objective testing is measured against a pre-determined standard, subjective testing needs a standard for comparative purposes. This is achieved by using experienced personnel whose knowledge of the product is such that they are able to detect small variations from the standard product.

A large proportion of bread sold is identified with a particular brand or producer and, in order to maintain customer loyalty, consistency is of importance. Where 'own-label' bread is being produced under contract, often for a large retailer, tight specifications are linked to the contract. Failure to meet these specifications can result in considerable financial loss due to rejection by the client, and the contract may be forfeited.

☐ **Machinery maintenance**

Throughout the system a great deal of specialized machinery is used. The success of the operation in terms of quality and profitability are strongly related to obtaining optimum performance from this machinery.

In the larger, more mechanized parts of the system, i.e. the flour mills and the plant bakeries, maintenance and checking the settings of machinery are routine procedures. This relates to the concept of a quality maintenance programme discussed previously. However, data from the quality control system on the products is used as an early indicator that adjustment is necessary. For instance, if significant numbers of dough pieces were rejected it would suggest that adjustment should be made to the dough divider.

■ Conclusion

This case study has attempted to give a brief insight into some of the most important aspects of bread production in the United Kingdom. Most of the considerations are not unique to bread and could equally apply to a number of manufactured foodstuffs. Indeed, the principles, although of course not the details, apply much more widely in other manufacturing industries.

3.3

The Compact Disc

F. Guterl

[This is an abridged version of the original article, which tells of the race to produce a commercial CD player for the mass market. We have replaced some of the more technical passages by a few sentences in square brackets in an effort to focus on the overall story.]

In 1969, it seemed a far-fetched notion at best, but the Dutch physicist couldn't get it out of his mind. One day he told a friend and colleague at the Philips plant in Eindhoven, the Netherlands, how RCA engineers in the United States had found a way to copy holograms inexpensively by impressing the refraction patterns on a nickel sheet that becomes a mould for pressing out the copies.

Physicist Klaas Compaan reckoned picture frames could be recorded in a similar microscopic way and reproduced on discs the size of a phonograph record. Once the images were on the disc, he suggested to Piet Kramer, then head of Philips's optical research laboratory, they could be projected onto a screen not merely in order, from first to last, but in any sequence whatever. The technique could prove an ideal training and education tool.

In that way was born the idea that grew to be compact-disc recording.

Kramer was intrigued. For a year and a half, Compaan spent much of his time in Kramer's laboratory working informally on the development of a prototype disc. Hardly anybody else, either within the company or outside it, then knew or cared about the work. It took 15 years for it to grow into a huge project involving hundreds of engineers at Philips and elsewhere in Europe and Japan. In 1983 it led to the introduction of one of the most successful consumer projects of all time: the CD digital audio system.

Although the idea that so struck Compaan was, of course, a long way removed from the end result, many of the principles he envisioned survived the long and tortuous road from research to development to product. His inexpensive, mould-pressed disc stayed alive, but instead of actual visual images, the discs were impressed with tiny dimple-like marks that represent digital bits; and instead of pictures the discs can reproduce any kind of sound with the highest fidelity imaginable.

The overwhelming commercial success of CD reproduction geared up a multibillion dollar industry around optical information and ploughed ground for a whole crop of optical-disc technologies that now promise to revolutionize data storage. CD-ROMs have already increased computers' capacity to hold recorded data; interactive CD-ROMs coupled with expert systems are expected to make it easier to handle data; erasable and write-once optical discs will be augmenting, in some applications perhaps even replacing, magnetic hard and floppy disks; and in the last few years, Compaan's original idea has been catching on in interactive video.

So at Philips Gloeilampenfabrieken NV, what started as an informal, part-time project was handed from one division to another, finding advocates all along the way. The CD's development stands as testament to technological creativity in a big corporation at a time when small entrepreneurs receive most of the credit for innovativeness.

It was engineering on the grand scale, the speedy development of an infant technology — including new material for the optical discs, a solid-state laser to read digital information, and optical servo-mechanisms and electronics for tracking and error correction — all packed into a rugged, portable, inexpensive unit. The project demanded cooperation not only among different groups within Philips, but also even between Philips and some of its competitors. Much of the final product design took place in collaboration with Sony Corporation, which participated with Philips in drawing up a standard CD format. [. . .] However, it took seven years — 1969 to 1976 — for optical-disc reproduction to take root.

[. . .]

By September 1971, Kramer, Compaan, and a handful of others had assembled a prototype that could read a black-and-white video signal off a spinning glass disc. In December, they had it up-and-running for senior managers at Philips; by July 1972, a colour prototype was demonstrated publicly.

■ The switch to sound

After the July 1972 demonstration, however, Philips began to consider putting sound, rather than video on the discs. The main attraction, it was thought at the time, was that the 12-inch (305 mm) discs would hold up to 48 hours of music. But Lou F. Ottens, director of product development for Philips's musical equipment division, put an end to any talk of an optical long-playing disc.

Ottens had developed the cassette-tape cartridge that swept through consumer magnetic-tape recording of the 1960s. He had thus had plenty of experience with the recording industry and no illusions that it would readily embrace a new medium. For one thing, he was convinced that the recording companies would consider 48 hours of music unmarketable; for another, he knew that any new medium would have to offer a dramatic improvement over existing vinyl records. Borrowing his own term for the by then commonplace cassette, he dubbed the proposed medium 'compact disc' – and decreed that the discs were to be no larger than a fair-sized beer coaster.

Ottens was encouraged to invest in the product by early reports that a solid-state laser, then under research at Philips, might be available for use [as a reader of information off the surface of the disc] by the late 1970s or early 1980s. At the time [helium-neon gas lasers were available but were] too bulky and expensive for an audio product. In addition, US regulatory authorities were resistant about the gas laser and were demanding what Ottens considered excessive shielding. Furthermore, a solid-state laser would require fewer parts, which would presumably bring down the system's price to where consumers could afford it.

Ottens assigned two engineers to pre-development work on the CD, based on Kramer's video prototype. They experimented with a helium-neon laser and a 305 mm LaserVision disc, using only the inner part of the disc – within a 70 mm diameter.

It was soon evident that FM was inadequate for CDs. [. . .] 'We needed to be better than an LP', said Ottens. [. . .] After experiments with a few analog codes, Ottens concluded that only a digital code would do.

■ Digital novelty

It is easy in the late 1980s to forget how recent is the proliferation of microprocessors and digital circuitry. In 1974, only three years or so after the first microprocessors appeared, designing a digital consumer product was risky, to say the least. George T. de Kruiff became Otten's manager of audio-product development in June 1974. He was amazed to find no digital-circuit specialists in the audio department and to see engineers still laying out printed circuit boards with strips of duct tape.

De Kruiff had only just begun to recruit new engineers and to buy computer-aided design tools when Ottens announced at a managers' meeting that the CD project was considering going digital. 'The decision was discussed quite a bit at the management level,' said de Kruiff. 'There was some doubt about whether the perfect audio quality justified a whole new medium. But Ottens believed strongly in the product and pushed it.'

Within a few months, Otten's engineers had rigged up a digital system [which sampled the audio signal] about 44 000 times a second – twice the maximum audible frequency of 20 000 hertz [. . .] and encoded it as a train of square pulses. [On replaying the samples through a 20 000 hertz filter the pulses could be converted back to the original audio signal, in much the same way that a sequence of still TV pictures, when replayed at a faster rate than the eye can follow, appear as a continuous movement. The sample amplitudes were represented on the disc by the lengths of dimples impressed on the surface of the disc. The light from the] helium-neon laser was reflected off the dimples to photodetectors [to reproduce the digital signal, which was then connected] in turn to a digital-to-analog converter.

The sound, clearer by far than either that produced by an FM system or that of an LP, was just what Ottens was looking for. With a 14-bit digital-to-analog converter, the best available at the time, the system's dynamic range was around 80 decibels, well above the 50–60 dB of LPs.

There was one serious snag, however. The digital system proved far more sensitive to errors than is FM, and the slightest scratch or dust particle on the disc would obliterate hundreds of bits and cause a blast like a cannon shot or a thunderclap. The system would require extensive decoding and error correction by hundreds of thousands of components, which at [integrated circuit] IC technology's infant state seemed likely to prove too expensive.

But between the thunderclaps and cannon shots flowed sound of unsurpassed clarity, and Ottens kept on with the project. A Philips management shuffle in 1976 made the CD project even more popular politically, Ottens said, and he argued anew that the risk was worth taking, prophesying that by the time a CD unit was ready for the factory [integrated circuits] would have progressed to the point of providing a way to implement the digital electronics inexpensively.

At that point the project changed radically. With the project's fundamentals defined, the goal was now to work out the details, reduce the system to tabletop size and cut the manufacturing cost to 150 Dutch guilders — at the time equivalent to about $50.

The CD system was a delicate balance of very fine optical components, where the smallest change in any one would affect the rest. Since each component involved virtually untried technology, but had nevertheless to be made as inexpensively as possible, engineering was especially difficult. However superior the Philips engineers thought their optical technology was likely to be, the phonograph had been refined over decades and had the entire recording industry to support it.

The CD had to produce sound of a level hitherto available only in professional systems costing thousands of dollars, yet it had to be inexpensive and sturdy enough to be bought and survive use by ordinary consumers. That market pressure placed high demands on the system's quality, reliability and cost.

The task was made even more complex by the need to enlist cooperation from such outsiders as the recording companies and even competing system manufacturers. Philips recognized that it would be of paramount importance to join with other manufacturers in negotiating a standard before introducing the actual product. The company made what was perhaps the most important procedural decision in the entire development of the product: it would demonstrate a prototype before other manufacturers several years before going into production.

It was a controversial and risky decision. For one thing, Philips engineers felt that they clearly had the technological edge on their competitors — Matsushita, Sony, RCA, and others. A demonstration would tip their hand on some key design features. And even if the demonstration did not give away the store, surely collaboration by design teams would. The alternative, as Philips saw it, would be to beat the competition to the marketplace only to see other companies jump in with incompatible designs. On the other hand, a standard would make it more likely that the CD would be accepted by the software companies and consumers.

Dozens of research and product-development engineers were put to the task, with Joop Sinjou as project leader. The project included setting up a program to break the product down into its various technologies and improve each one, devising better methods, eliminating redundant components, then integrating them back into a whole product. In the research laboratories, Marino Carasso led a parallel team that concentrated on the more intractable scientific obstacles, most particularly the optical system.

Even before the prototype could be readied, Philips had to persuade a record manufacturer to develop an inexpensive disc material with the proper optical properties and that would not warp. The obvious choice was Polygram Record Service GmbH, Hannover, West Germany, which had worked with Philips in 50–50 joint ventures on projects such as the cassette tape. Polygram was at first cool to the idea, reasoning that vinyl LPs came as close to perfection as consumers would ever want.

It was not until 1978, when Philips demonstrated a prototype to Polygram's management, that the disc's development began in earnest. 'They had loaned us the master tape we used in cutting the CD,' said de Kruiff. 'When we demonstrated it, they listened and listened and thought they heard a rumbling sound. We thought that was impossible, of course. But it turned out there had been sound leaking into their recording studio from the outside — street noises — and you could hear that on the CD! They had to go back and reinsulate their studio, but they were enthusiastic about the CD from then on.'

■ Developing the discs

Polygram had the equivalent of about $150 000 and three months to develop the disc. It experimented with a plastic developed for the LaserVision, but its tendency to absorb moisture made it unsuitable for the small CDs. Since the clear coating of plastic that protected the dimples kept moisture off one side, the disc tended to warp. With LaserVision the problem was mitigated by gluing two discs together to form one rigid double-sided disc, but that would have been too expensive for a CD, and it was felt that for consumers to have to turn over the disc after each half-hour of music would be unacceptable.

Polygram created a polycarbonate plastic that seemed to do the trick – it was rigid enough and did not warp – but it had a troublesome tendency to alter slightly the incidental light's polarity as a result of birefringence – the splitting of the incident light into two rays. Since the optical system was at the time using a polarized beam splitter for tracking, Polygram engineers tried again but failed to find a better solution.

Back at Philips, meanwhile, the engineers were tackling other problems. One was to choose the type of laser, and after examining a number of alternatives they decided that the gallium arsenide [GaAs] diode laser was the only way to go. 'It was a rather brave decision,' said Jacques P.J. Heemskerk, an optical specialist who moved from research to the product-design team. 'At the time Philips used a few diode lasers in telecommunications applications – but these numbered in the tens.' Philips nevertheless decided to use the GaAs laser diode in a product that was to run to hundreds of thousands of units.

In the end, of course, the decision turned out to be correct. But over the next several years, with the 1983 product introduction date looming closer, Philips searched for a suitable laser diode – those from several companies either proving too noisy or having unacceptably short lifetimes. Philips tried to make the part in-house but was confronted by low production yields. In 1981, Sharp Corporation, not known previously for any laser expertise, unexpectedly came out with a long-life product and agreed to manufacture a laser to Philips's specifications.

▉ Simplifying the lens

While the components people wrestled with the laser, the optical-system designers tried to reduce the number of parts so as to cut manufacturing cost and improve reliability. The first thing to be revised was the objective lens, purchased from a microscope manufacturer. While it focused [the laser beam] accurately, [. . .] its four separate spherical lenses made it far too expensive.

Philips engineers dropped the requirement that the lens must be immune to chromatic aberration, reasoning that only one colour of light would be used; next, they designed an aspherical lens that would do the work of all four spherical components. Since most lenses are made by circular grinding and polishing of a rough cast, their surfaces must be a segment of a sphere, and often several must be used to get the desired result. With high-precision moulds, however, there is no need for grinding and polishing, and a lens can be made to any shape. Philips's laboratories were already making such high-precision moulds.

[. . .]

[Two problems with the optical system remained: how to keep the beam on the dimple track and how to keep it focused exactly on the disc's surface. Servomechanisms were developed for both, in which reflected light from a split beam was detected by photodiodes. If the beam moved off the track, or if it got out of focus, the photodiodes would activate the necessary adjustments to bring the beam back on track and in focus again.]

An early dispute over the design of the error-correction circuitry focused on whether the CD should play from the disc's rim inward, or from near the spindle hole outward. The product engineers wanted to play from the rim, reasoning that the larger radius of the outer tracks would mean more spread out dimples there and so fewer errors in the music. Some older engineers, however, could recall early phonograph users complaining that the tone-arm kept dropping off the record's rim, and the trouble that resulted in keeping variations in the record size to the minimum.

The decision was made to have the laser pickup begin near the disc's centre; to preserve the sound's quality, the motor spinning the disc would vary its speed, gradually slowing as the pickup advanced toward the rim. The linear velocity [of the dimples past the laser beam] would thus remain constant, [allowing constant dimple sizes, and no crowding of the track near the centre].

■ Correcting for errors

But when it came to working out the actual error-correction code, Philips came up against a wall despite its experience in signal processing. [The requirement was to devise a way of adding extra digits to the coded digital signal, in such a way that the decoding circuits could 'infer' what was missing when scratches or other imperfections damaged the dimple track. Just as people can often infer a missing word in another's speech from the context of the words around it, so the added digits had to give sufficient information about missing or damaged digits in the CD track for the electronic circuits to replace them accurately − with negligible distortion. One way of doing this is referred to as a 'convolution algorithm'.] The CD system needed such high reliability and such finicky digital electronics that for several years it seemed as though any solution using this method would entail unacceptable compromises.

[. . .]

But towards the March 1979 deadline, the product-engineering group was still having problems with the convolution algorithm. 'It was great when it caught an error because it would completely eliminate it,' recalls Jos Timmermans, then with the product-engineering group and now the CD group's chief engineer for electronic design. 'But when an error came along that it couldn't catch, the redundancy in the code just made it worse, and the error would show up over and over again in the decoded signal.'

[. . .]

At the demonstrations Philips engineers had to go with what they had. The electronics were wired onto printed circuit boards, but it still took three months to synchronize the circuits' timing. After working through the night and into the morning of the demonstration, however, the product-engineering group brought their managers into the conference room.

There a compact-disc player sat on a table covered by a floor-length green cloth. A small disc was shown around the room – handled very carefully so as not to smear it with fingerprints that might elude the as yet unfinished error-correcting algorithm. Once the music was played and the managers were suitably impressed, the cloth was pulled aside to show the rows of printed circuit boards with their tangle of interconnecting wires, all later to be packed into four or five ICs [integrated circuits].

Weeks later, the prototype was flown to Japan and demonstrated before several manufacturers; this time, however, a cloth also covered the mechanism on the table. The first visitors were from Matsushita, which turned down the Philips offer to work on a standard. Next came Akio Morita, Sony's chairman, and he and Philips chairman Cornelius van der Klugt eventually agreed to work together.

■ Information exchange

Sony and Philips engineers began monthly meetings in Tokyo and Eindhoven where each group would present proposals for details in the CD system. There were hundreds of details to be considered: the size and shape of the dimples, the data format, the configuration and size of the discs, [. . .] the sampling frequency [and so on].

[. . .] Philips engineers involved in the negotiations recall that the Sony and Philips team were complementary in their strengths. Philips had approached the project from a theoretical point of view, starting from the company's optical experience and applying that to sound recording. The Dutch team was thus farther ahead on the optical side of the design, and even though Philips had derived experience in digital coding from its work in telecommunications, Sony had been making professional audio equipment for years and brought a wealth of knowledge about digital sampling techniques.

Sony was a leader in magnetic-tape recording and the use of pulse-code modulation, digital conversion of analog signals, sampling, filtering, and other techniques that had a big impact on the CD player's electronics. Sony also had practical expertise in error correction that Philips engineers were struggling to learn about. [. . .] Between the monthly discussions, Philips engineers would work on the Sony proposals, often slightly improving the algorithms within the overall structure of the Sony code. 'We kept going back and forth on the codes,' said Peter W. Bogels [. . .] 'until we had something much better than either side started out with.' [The method finally proposed by Sony, called the 'cross-interleaved Reed–Soloman code' was able to correct burst errors up to 2400 bits long, which corresponded to a 2 mm scratch!]

Sony also insisted on a 16-bit digital-to-analog converter, rather than Philips's 14-bit component. The Philips team argued that the extra 2 bits per sample would call for a bigger disc without adding substantially to the quality of the sound. Listening trials proved them wrong, and the disc was increased to 120 mm from 115 mm. Later, Sony's Morita insisted on another increase to allow 75 instead of 60 minutes of music — enough to accommodate all of Beethoven's Ninth Symphony, a favourite of his wife.

The exchanges with Sony ended in 1981, after about a year of meetings. During that time, Philips engineers say, Sony had learned a great deal about Philips's optical tracking design that had to be revealed during the error-correction work to explain some of Philips's test results. That gave away much of the lead in the product's development, but Philips saved time in learning how to code the signal. The final product was a hybrid of the two teams' ideas. [. . .]

■ The final lap

All that was left was to work out the actual circuit design that would implement the standard format. Philips and Sony agreed that if either designed a circuit that did not work out, the other would offer theirs, but the provision was never used. Philips and Sony engineers had no further communication, and the race was on.

Philips engineers had to redesign completely the system's electronics, make a new prototype, spend three months synchronizing and testing it, and then make the ICs. Although the process was straightforward, Philips had had little experience in digital electronics and ICs and the work dragged. It took 18 months to reduce the jumble of boards and wires to five chips.

Since Sony was an old digital hand, it was able to overcome its rival's lead in the optics, and it beat Philips to market by a month. 'We thought we had a six-month lead, no problem,' said Timmermans. 'We were a bit disappointed. Later we asked them how they made the ICs so fast, and they asked us how we were so slow.'

Sony's swiftness in the product design was also due to its different design philosophy, Timmermans said. 'Throughout their product development, Sony had the attitude that they would build something and then work on improving its weak points,' he said. 'We would take a more academic point of view, spend time studying and developing the theory, and so on. We tried to do more in one step. We had a plan as to how we were going to improve the product and I don't think Sony did. We had digital filtering in our first product and Sony didn't. Sony had the attitude, I think, of getting the product out the door quickly and worrying about improving it later.'

Once the design was finished, the Philips design team moved to the company's plant in Hassalt, Belgium, to iron out last-minute manufacturing problems. In 1983, Philips made almost 100 000 units, each containing 20 ICs. Last year the company produced 2.5 million units, each with only six chips. 'We keep improving them,' says Timmermans, now chief engineer at the Hassalt plant. 'When we can make them with one IC, we'll probably stop.'

3.4

Evolution of Bridges

C.J. Burgoyne

Bridges are now so common that we take little notice of them; there are about 200 000 in Britain alone, with about two per mile on motorways and trunk roads. But bridges are essential to our economic well-being, something we only fully realize when one has to be closed for repairs. Their importance is also well known to the military; ever since Horatio saved Rome by defending the bridge over the Tiber, military commanders have variously sought to defend, capture or destroy bridges for strategic advantage.

To structural engineers, bridges represent the clearest expression of their art; they are usually unadorned, and the structural form is clearly seen. Big bridges are often working at the limit of what is possible, and there is a degree of national pride in having 'the world's longest span' in Britain, a record we are soon to lose.

Historically, bridges have changed in response to improvements in technology; materials have changed, understanding of structural behaviour has improved, and new manufacturing technologies have been developed. We shall follow these changes, in roughly chronological order, and look at the legacy we can see in the bridges left behind.

The earliest prehistoric bridges would have been beam bridges built from tree trunks; inevitably, none now remain. Stone was better, but very few stones can be cut into single pieces long enough to be useful to span a gap, while at the same time being thin enough to be lifted into position. There were never many examples like the clapper bridges on Dartmoor, which are believed to be several thousand years old (Figure 1).

The most significant structural development in the ancient world was the invention of the arch (see Box 1). The first arches are believed to have been built in the Babylonian and Sumerian civilizations centred around the Tigris and Euphrates rivers; without suitable stone, and with insufficient trees, they needed some structural form that could be built with clay bricks. Surprisingly, the Greeks did not know about the arch, which explains why the columns on their temples are so close together; the gap had to be small enough to be bridged by a single stone. The Romans, however, made extensive use of arches,

199

Box 1   Structural form 1: The arch

The weight of the material or loads above the arch causes a line of thrust which passes through the ring of blocks (voussoirs) which make up the arch. The exact line that this thrust follows does not matter particularly. Provided it lies within the thickness of the ring, then the arch will not fail by instability, and if it lies within the middle third of the thickness, the arch will not crack. The stresses are normally well within the capacity of the stone or concrete from which it is constructed, so failure by crushing is very unlikely.

The most important aspect of arch behaviour is that the arch thrusts against the abutment, which must be able to resist that thrust; the shallower the arch, the larger the thrust. The thrust can be resisted either by rock foundations, by the weight of a solid abutment block, by the thrust from a neighbouring arch, or by tie rods between the arch foundations.

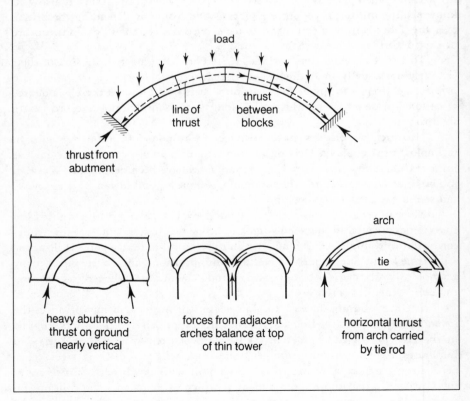

Figure 1 A clapper bridge at Postbridge on Dartmoor.

and many examples remain to this day (see Figure 2), although there are none in Britain that can be attributed to them with certainty. The Romans also made extensive use of timber bridges, with beams resting on trestles built on piles driven into the river bed. None of these remain, most having been replaced by more permanent structures on the same alignment.

In certain parts of the world, cantilever bridges were developed. If timbers are extended out from the bank of a river, they will eventually topple into the

Figure 2 An early Roman arch bridge.

water, but this can be prevented if extra weight is added on the shoreward side. By repeating this several times, spans in excess of the longest timbers can be bridged. In other places, rope bridges made from natural fibres were developed, usually to cross narrow gorges.

During the Middle Ages, most bridges would have been built from timber, or from masonry arches. One of the major difficulties with arches is that they need very firm foundations, and are not stable until complete. Massive masonry piers are needed, and the voussoirs need to be supported on timber scaffolding until the ring is complete. Then comes the delicate operation of removing the scaffolding and transferring the load to the arch; only at this stage would the mediaeval mason have known whether he had got the shape right, and whether the supports were sufficiently rigid. There was no theory about the best shape to be adopted, only masons' folklore, and many would have failed immediately (if the shape was wrong) or within a few weeks (if the abutments started to move).

If a long bridge was being built, with several arch spans, the thrusts from the arches on each side of a pier would balance when the bridge was complete, which allowed the piers to be kept fairly narrow. However, the mason would have needed to support several spans at one time to prevent the piers being unbalanced during construction. For this reason, many old masonry viaducts have massive piers which are capable of resisting the force from the arch on one side only, so they should really be regarded as a series of separate bridges placed end-to-end. The size of the piers would restrict the flow of water under the bridge, which could lead to problems of scour caused by water washing away the foundations. The water on the upstream side of the old London Bridge could be several feet higher than on the downstream side, making passage in a boat very hazardous.

In the 17th and 18th centuries, intellectual thought turned to the way loads were carried in structures. In 1675, Hooke published a Latin anagram, which when translated reads 'As hangs the flexible line, so but inverted will stand the rigid arch', which for the first time explained how the shape of the line of thrust within an arch ring could be visualized (Figure 3). But it was not until 1773 that Coulomb correctly established the principle that governs the stability of arches and the relationship between the shape of the arch and the line of thrust.

Coulomb and others also worked on the behaviour of beams, but it was Navier in 1826 who correctly analysed the general behaviour of a beam in bending, relating the internal stresses to the externally applied load. The mathematician Euler had already studied the deflected shapes taken by elastic beams under various loadings (without considering stresses), and deduced (in 1744) that beams could buckle when subjected to purely compressive loadings. Other developments built on these basic principles, most notably in theories relating to the behaviour of trusses, so that by the time St Venant solved the problem of torsion in beams in 1855, the principles which govern the behaviour of bridges were largely known.

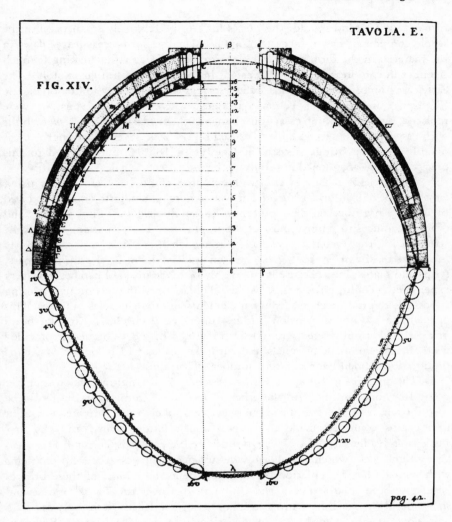

Figure 3   An illustration of the line of thrust (as a chain) within an arch ring; from an analysis by Poleni of a section of the dome of St Peter's in Rome. The thrust line was determined experimentally by loading a flexible string with unequal weights, each weight being proportional to that of a segment of the line, and due allowance being made for the weight of the lantern. [. . .] The thrust line found in this way does in fact lie within the thickness of the dome of St Peter's; the figure also shows that a uniformly loaded string would produce an equivalent thrust line passing outside the masonry.

These theoretical developments produced relatively little change in bridge building until the industrial revolution brought about the development of new materials. *Cast iron* is produced by pouring molten pig iron, made in a blast furnace, into a sand mould; the pieces have to be relatively small because the whole component has to be cast in one go. The casting process leaves flaws in the material in the form of air bubbles or pieces of slag, thus making it much stronger in compression than in tension; in this way it is similar to stone. The early cast iron bridges were thus arches, such as the Ironbridge in Coalbrookdale (Figure 4) and many of Telford's bridges, where the stone voussoirs were replaced by interlocking cast iron elements. Many beam bridges were built with cast iron, but several failures caused by the lack of tensile strength, most notably the Dee Bridge disaster at Chester in 1847 (Figure 5), led to this practice being abandoned for railway bridges.

The lack of tensile strength of cast iron could be overcome by using another type of iron, *wrought iron*. This was a more refined form of iron produced by melting pig iron and then 'puddling' it in a furnace with more iron ore and other substances to remove most of the impurities. The processing altered the structure of the material, producing much smaller flaws and significantly increased tensile strength. Wrought iron was available in the form of bars, strips and plates, but, because it could only be made in small batches, was very expensive. It could, however, be worked by blacksmithing techniques, making it possible to produce chains from bar elements and to drill holes in plates. This opened the way for suspension bridges supported from chains, and for beam bridges made from riveted wrought iron plates. These developments coincided with the development of canals and turnpike roads, which for the first time produced a demand for a large number of bridges.

The suspension bridge (see Box 2) relies on a hanging tension member going from one abutment to the other. Ropes made from natural fibre can be used for relatively short bridges, but they are unlikely to be durable. Wrought iron chains could be made to support much longer spans, and the first suspension bridges in the UK were built in 1819. Early suspension bridges had a very light deck, which did not provide sufficient stiffness to prevent problems with wind-induced oscillations or excessive deflection. Most of these bridges have now been modified, but the Union Bridge over the Tweed and several small footbridges still have unstiffened decks which deflect alarmingly as a load passes over them. Telford's suspension bridge over the Menai Straits (Figure 6) suffered similar problems, with the deck having to be replaced soon after it was built, and again in the 1940s.

Iron can also be produced in the form of wire which, if drawn through a series of dies to successively reduce its diameter, can have exceptionally high strength. This development led to the construction of suspension bridge cables from bundles of drawn wire, rather than chains, which allowed much larger spans to be constructed. The foremost exponent of this form of construction was probably Roebling, in the United States, who even built a railway suspension bridge near the Niagara Falls.

Beam bridges (see Box 3) are subjected to both compressive and tensile stresses, and if long spans are required, methods must be available to join components together. Compressive forces are easy to transmit, simply by

Figure 4   Ironbridge, the cast iron bridge at Coalbrookdale, UK.

SCENE OF THE LATE RAILWAY ACCIDENT, AT CHESTER.—DILAPIDATED SPAN OF THE DEE BRIDGE.

Figure 5   An engraving of the Dee Bridge disaster (1847), UK; it collapsed because a tension member broke.

Box 2   Structural form 2: The suspension bridge

If a series of weights are hung from a cable which is tied to rigid supports at the abutment, it takes up a curved profile known as a *catenary*. If a deck is suspended from the cable by means of hangers, a level roadway can be formed. A suspension bridge built in this way is very strong, but at the same time very flexible. The bridge can deflect significantly under heavy loads, and can vibrate under wind loadings. It must thus be stiffened by having either a heavy truss deck, a stiff box beam deck, or by inclined hangers.

The profile taken up by the suspension cable is the inverse of the line of thrust in an arch. In the same way that the arch pushes on the abutment, so the suspension bridge pulls on the abutment, which must be sufficiently massive that it can resist the cable forces.

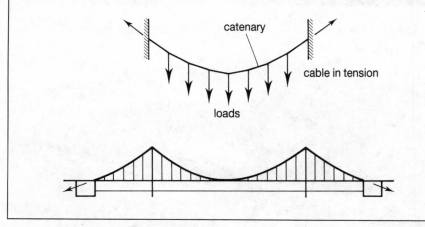

---

butting one element up against the other, but tensile forces cause more problems. The ability to fabricate wrought iron by riveting plates together meant that larger spans carrying heavy loads started to become possible for the first time. This was put to good use with the advent and rapid expansion of the railways, which required a large number of bridges, not just over rivers, but also over roads and valleys.

Brunel and the Stephensons, father and son, were the foremost railway engineers in Britain, and indeed were influential overseas. Most railway bridges were built of brick arches, but some spans became larger and flatter, such as Brunel's Maidenhead Bridge, which pushed conventional ideas to their limit. In the early years of the railways there were a number of notable timber bridges, generally on the more remote railways, such as the trestle viaducts built by

Box 3   Structural form 3: The beam

If a beam rests between two supports, with a load on the top, it will take up a slightly curved form, with the bottom of the beam getting longer, and the top getting shorter. Material in the bottom of the beam is thus in tension, and material in the top is in compression. If the beam is *cantilevered* from a wall, the situation is reversed, with tension at the top, and compression at the bottom.

If the bridge is to pass over several supports, each span can be made from a separate beam, but it is more efficient to make the beam *continuous*. Tension stresses are caused in the top of the beam over the supports and in the bottom of the beam at mid-span, with compressive stresses in the opposite faces. There are no joints (which are difficult to maintain) and the beam can generally be made lighter and cheaper than the *simply-supported* alternative.

Beams have to be made from material which can resist both tension and compression, or by a combination of materials, one of which is good in tension, the other good in compression.

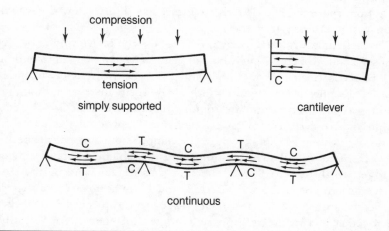

Brunel in Devon and Cornwall. Most were later replaced by more permanent structures.

The need to cross major rivers where navigation spans were required led to the biggest developments in structural form. The most important of these were over the Tyne, the Menai Straits, and the Tamar, all of which produced different innovative solutions. To cross the Tyne, George Stephenson produced a design using a *tied arch*, where the thrust of the arch is carried, not by the piers, but by tension rods tying the ends together. The bridge has two decks,

Figure 6   Two different designs in the Menai Straits in North Wales: the Menai Suspension Bridge (foreground) and the Britannia Tubular Bridge.

one for the railway supported above the arch, and a lower one for road vehicles hanging from the arch. The tie rods are incorporated just below the road deck, and can easily be seen from below.

The Britannia Bridge (Figure 6) built across the Menai Straits by George Stephenson's son, Robert, is innovative in many ways. He wanted to build large arches, but was prevented by the Admiralty who insisted on maintaining navigation clearances over the full width of the span. Instead, a wrought iron box beam was used, with the trains running on rails inside the box. One flange of the box has to carry compressive forces, but Stephenson was aware of the problems of buckling of thin plates, so the compression flange of the beam was made with stiffening cells which eliminated the problem. He built a model of the bridge, which he tested before starting construction, and built a simpler single-span prototype at Conwy. The individual box beams were fabricated on shore, floated into position, and then jacked up the piers, where they were joined together end-to-end to produce a single continuous beam. Finally, he jacked the bearings at the pier positions to induce loads in the beam which reduce the effects of the beam's weight, a technique which is regarded as advanced, even today. Stephenson had made the towers high enough to support chains which he would have used to provide additional strengthening, but in the event they were not needed. Paradoxically, when the boxes were damaged beyond repair by a fire started by vandals in 1970, they were replaced by steel arches similar to Stephenson's original design, but with sufficient strength to carry a roadway to relieve Telford's nearby suspension bridge.

At the same time as the Britannia Bridge was under construction, Brunel was tackling a similar sized structure at Saltash, where he had to cross the River

Figure 7   Brunel's bridge at Saltash; adjacent is a modern
suspension bridge.

Tamar (Figure 7). He adopted an equally innovative approach, with a hybrid
structure consisting of wrought iron arch tubes and a suspension chain. The
load on the bridge has to be shared between these two components so that the
outward thrust of the arch is balanced by the inward pull of the chains; even
with today's sophisticated methods of analysis that would be no mean
achievement. The bridge was lighter than the Britannia Bridge, and is still in
use, but significantly has never been copied.

Similar achievements were made elsewhere. Wrought iron trusses (Box 4)
and arches were in widespread use, the most notable exponent being Eiffel who
built several large structures on the continent.

The next innovation was the development, in 1856 by Bessemer, of ways
of producing steel economically in large quantities. Steel is basically iron, to
which a small amount of carbon has been added. This has the effect of changing
the crystal structure quite extensively and allows both tensile and compressive
stresses to be carried. Steel quickly replaced both cast iron and wrought iron in
bridge building, and opened the way for new developments.

A turning point in bridge design was the Tay Bridge disaster, which

Box 4   The truss bridge

A truss bridge behaves like a beam, but with most of the material removed to save weight. The top and bottom chords of the bridge carry the main tension and compression forces, but the diagonal elements transmit forces between the chords, and also brace the compression chord against buckling. The force in these diagonal elements can change from tension to compression as the load traverses the beam, and in larger trusses, the diagonals also need to be braced.

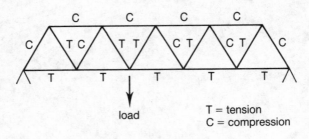

load

T = tension
C = compression

occurred in 1879. The bridge had been built of light, wrought-iron trusses, supported on cast-iron piers, with very little bracing across the structure. The loading induced by wind on the structure, especially when carrying a train, had not been properly understood, and 75 people were killed when the bridge collapsed during the passage of a train in a gale. The Forth Bridge (Figure 8) was built shortly afterwards using steel for the first time in a major bridge. It was designed with due allowance for wind loads and despite its solid appearance in a foreshortened view, it is seen to be a remarkably light structure when viewed from the side.

At about the same time as steel was first used for bridges, concrete started to be used as a structural material. The Romans knew about lime-based cements and they mixed these with crushed clay tiles to make a sort of concrete. In 1824, Aspdin produced a much stronger cement by fusing together limestone and clay, thus combining the two materials much more intimately. This cement reacts with water to produce a solid matrix, which can be used as a binder for aggregate (sand and gravel). Aspdin called his product Portland Cement, since the resulting concrete resembled Portland stone in colour and texture.

At first, concrete was considered as a means of moulding stone, rather than having to carve it. Osborne House, built on the Isle of Wight by Prince Albert, was constructed in this way. But engineers began to realize that they could cast a complete arch ring from concrete, without any joints; the first such

Figure 8 The Forth Railway Bridge over the Firth of Forth in Scotland, 1890, by Benjamin Baker. With two spans of 1710 feet, this steel cantilever bridge surpassed Brooklyn Bridge as the world's longest span. The steel structure rises 342 feet above the masonry piers. Although from a foreshortened view the bridge appears dense and massive, in profile it exhibits a surprising lightness.

bridge was built over the River Axe at Seaton in Devon in 1877. Concrete was developed too late for the majority of the railways but some later railway bridges such as Glenfinnan Viaduct in Scotland were built with mass concrete.

It was not long, however, before engineers started to think of concrete as a material in its own right, rather than as a replacement. The biggest problem with stone and concrete is that they have very little tensile strength, and are brittle, although they have high compressive strength. Steel rods have high tensile strength, but a tendency to buckle when in compression. This led to the development, by Hennebique of France in 1892, of *reinforced concrete*, where bars of steel are placed in the concrete to resist the tensile forces, leaving the concrete to carry the compressive forces. We thus have an ideal material for constructing beams.

In the early years of this century many reinforced concrete bridges were constructed. At first, many of them were of the open arch type, with a relatively thin reinforced concrete ring (Figure 9). This supported columns which in turn carried the roadway. The arch was primarily acting in compression, but high point loads induced through the columns would cause bending in the arch, which could now be resisted because of the reinforcement. Smaller structures were built as simple beams, or beams supporting a thin slab. The freedom to choose the shape of the concrete section, and at the same time to vary the strength by altering the amount of steel reinforcement, at last gave engineers the chance to explore a wide range of structural forms.

The steel bars in reinforced concrete are much stiffer than the surrounding concrete, which has very little tensile strength. This means that the concrete cracks before the steel carries any significant load. In properly designed structures, these cracks are invisible to the untrained eye, but they lead to a loss

Figure 9   A thin reinforced concrete arch bridge. The Heads of the Valley Taf Fechan Bridge, Wales, UK.

of stiffness of the beam as a whole. It had long been realized that these cracks could be elminated if the whole concrete beam were subjected to an overall compressive force, so that the effect of the loads merely reduced the original compression, rather than causing tension. Attempts were made, as early as the 1890s, to do this by tensioning reinforcing bars running through the concrete, but after a few weeks the cracks reappeared.

It was only when the French engineer Freyssinet realized that concrete creeps under load, in 1927, that *prestressed concrete* was developed. He realized that if very high strength wire was tensioned, and good quality concrete cast around it, a significant compression would be left in the concrete even after the creep had taken place. Prestressed concrete bridges are normally stronger and stiffer than the equivalent reinforced concrete structures, which means that they can be made lighter. Because they have no cracks, and are made from better quality concrete, they are also more durable. Virtually all concrete bridges with spans greater than about 15 m are now built from prestressed concrete, in a variety of structural forms (Figure 10).

Developments in structural theory had matched changes in materials, but there were great difficulties in doing the calculations. The behaviour of most structural elements can be described in terms of differential equations, but in only a very limited number of cases can these equations be solved analytically; most such solutions had been obtained by the early years of this century. However, these differential equations can be rearranged into sets of simultaneous equations for more general problems. The number of calculations required to solve a set of equations increases as the cube of the number of unknowns; when calculations are performed by hand, there is a significant chance of making an error, so very careful procedures are needed. There was thus a great premium on minimizing the number of equations to be solved at one time, and a number of techniques were developed to achieve this. Various methods of finding approximate solutions, many involving iterative relaxation methods, were adopted in the first half of this century, which allowed most designers to get at least a reasonable approximation to the behaviour of the structures they were designing.

Improvements in steels have also taken place in the 20th century, with higher strengths and weldable steels becoming commonly available. Welding, where two components are combined by melting the steel at the join by means of an electric arc, does away with the need for lap joints and is much less labour-intensive than the riveting process that it replaced. But it does require special steels and skilled operators, since the welding process upsets the crystal structure of the metal near the weld, which can lead to a brittle zone and the formation of cracks. This was first demonstrated during the Second World War

Figure 10 A prestressed concrete bridge by Freyssinet at Pont de Luzancy.

Figure 11   Box-girder bridge under construction, Avonmouth, UK.

when mass-produced Liberty ships started to break up when operating in cold waters. A number of bridge failures can also be ascribed to these effects. Welding is not usually a problem in factory conditions, but is more difficult to control on site, so hybrid methods of fabricating steel bridges are often used. Large components are welded in a workshop, and then joined together on site by high strength bolts.

Two bridge forms have come into common use in the second half of the 20th century; *box-girder* bridges, and *cable-stayed* bridges. Box-girder bridges are of cellular form, which makes them torsionally stiff, and are thus very good at distributing heavy loads placed on one side of the structure. They are economical in their use of material, but can be complex to construct (Figure 11). Steel boxes are often fabricated from flat plate components, stiffened to prevent buckling, and concrete box girders are also widely used.

Cable-stayed bridges (see Box 5) are often confused with suspension bridges, but behave in a very different way. They have steel or concrete box decks, supported by steel wire cables which radiate out from the top of a tower (Figure 12). They do not suffer the same oscillation problems as suspension bridges, and evolved from early hybrid structures, such as the Albert Bridge in London and the Brooklyn Bridge in New York, which are primarily suspension bridges but have stay cables to improve their stiffness. Most large span bridges (but not usually the very largest) are constructed in a cable-stayed form.

All of these recent advances have been made much easier by the advent of the computer, which has removed the problems associated with solving simultaneous equations. Before 1960, solving five simultaneous equations was a major exercise, whereas today many engineers use micro-computers which can solve thousands of equations within a few minutes. It is possible to analyse a structure to any degree of detail required; the problem is becoming one of *too much* information, from which the engineer needs judgement to extract the

Box 5   Cable-stayed bridges

The cables in these bridges act as supports for the deck structure. They serve to reduce the bending in the deck which can be made quite thin and therefore light. The beam acts as though it were continuous over a large number of spans, although the cables are extensible and do not act as rigid supports to the beam, which must be taken into account in design.

Cable-stayed bridges are normally governed by overall stiffness requirements, rather than strength conditions, and fatigue problems in the cables must be considered. They are much less susceptible to wind-induced oscillations than suspension bridges once complete, although they can be vulnerable to high winds during construction.

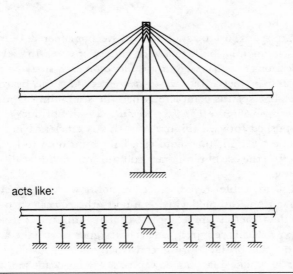

acts like:

---

relevant results. Future developments will almost certainly be in the field of *expert systems*, which will assist the designer in the preparation of designs and in the interpretation of the results of analyses.

In the 1990s, the engineer has a wide range of bridge types to choose from, but certain standard solutions are now apparent. Arches are rarely used, except where good foundations exist to resist the thrust and where the opening to be bridged is of suitable shape. For small spans, up to about 15 m, reinforced concrete beams or slabs would normally be used, but beyond that, prestressed concrete becomes most economic. For spans up to about 25 m, which covers the large majority of structures, bridges built from precast prestressed concrete beams, lifted into place and then topped with a reinforced concrete slab, are usually used.

Figure 12   The cable-stayed bridge at Dartford, Kent, UK.

For medium spans, up to about 150 m, steel or concrete beam bridges are usually the most economic, built in a wide variety of ways. In *balanced cantilever* construction, sections of equal weight are added on each side of a pier until the two halves of a span meet. In *span-by-span* methods, the beam is built on scaffolding until one span is complete, when the scaffolding is moved forward to build the next span (Figure 13). *Incrementally launched* bridges are built at one abutment and pushed forward (Figure 14); this is efficient since construction takes place at one location, but requires a launching nose to be added to the structure, and often the use of temporary columns to provide additional support during launching.

Beyond 150 m, cable-stayed structures are now the most common. The deck is built out from the pier which supports the tower, with cables being added as each few metres are completed. This minimizes the amount of temporary support that is required, but the designer must take account of the varying loads on the bridge during construction, and also the quite large deflections that will take place as new pieces of the beam are added. Beam elements and cable lengths have to be adjusted so that the bridge ends up in the correct shape when complete, a problem that is only really solvable with the aid of modern computer power.

The very largest bridges are still built as suspension bridges, but only for spans that exceed about 600 m (Figure 15). Knowledge of structural behaviour and methods of analysis have improved to the extent that some of these bridges now carry railways, as well as roads. The high concentrated loads from heavy trains, particularly from the engines, were long seen as causing unacceptable deformations.

What are the problems with existing bridges? Bridges are often used at the limit of what is economically possible, so it is not surprising that a small number suffer from problems. Around 1970, four major steel box-girder bridges collapsed, three of them during erection, when the effects of interaction between buckling and yielding of the steel were not sufficiently understood.

Figure 13   'Span-by-span' construction method.

Other problems relate to durability. If water is allowed to penetrate a structure, steel can corrode, particularly if the water contains de-icing salt that has been spread on the roads. This is clearly a problem in steel structures, but the problem is usually visible and can be prevented by regular maintenance. A more insidious problem exists in concrete bridges where the corrosion of reinforcing or prestressing steel is normally prevented by the highly alkali environment provided by the concrete. However, the alkalinity can be destroyed by the penetration of atmospheric carbon dioxide or chlorides from sea spray or de-icing salts. When this occurs, reinforcement can rust, causing the outer layers of concrete to spall off, or the prestressing wires can fail, which leads to collapse of the structure. There has been only one such collapse in the UK to date, but these faults are almost impossible to detect from the outside.

What of the future? A number of new materials have been developed with the potential to overcome the durability problem, and at the same time extend the size of structures that are possible. These are all in the form of fibres, made from glass, carbon or aramids, and have strengths and stiffnesses that compare well with those of steel. These fibres can be aggregated into ropes or made into composite bars with epoxy, either of which can then be used as prestressing tendons or stay cables. Bridges have been built with all these new materials in the form of prestressing tendons, and their use is likely to increase if the durability of steel proves to be a major problem.

The real advance will be made when the largest suspension bridges are constructed from these new materials. The current largest span is the Humber Bridge at 1.4 km, but there is a 1.6 km span under construction across the

Figure 14 'Incrementally launched' bridge construction method, Dornoch Firth
Bridge, Scotland.

Great Belt in Denmark, and a 2.0 km span across the Inland Sea in Japan
(Figure 15). These all use steel cables, but steel is very heavy, and it is generally
believed that a 2.5 km span is the largest that could be constructed with steel.
Carbon and aramid fibres have a much higher strength/weight ratio, which
would allow longer spans to be constructed. The Messina Bridge between Italy
and Sicily would need a 3.5 km span, and a series of 4.5 km spans with aramid
ropes were proposed for the Channel crossing. The largest span proposed to date
is for an 8.4 km span across the Strait of Gibraltar, supported by carbon fibres.

Whether any of these bridges will ever be built remains conjecture. What
is clear is that bridge forms will continue to evolve as new materials are
produced, new construction techniques developed, and as we improve our
understanding of the way these structures behave.

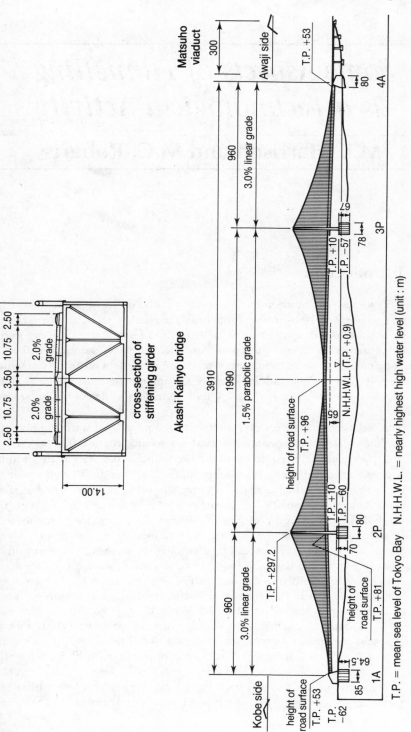

Figure 15 The proposed Akashi Bridge (suspension) in Japan.

3.5

Some Aspects of Tunnelling as a Technological Activity

M.E. Harrison and M.C. Roberts

■ Introduction

Public awareness of tunnels tends to be mainly of those associated with transport systems. High-profile tunnelling activities, such as the Channel tunnel, or accidents that occur in tunnels, tend to focus attention temporarily on their construction or use. But apart from such instances, tunnels tend to be taken for granted. A little thought reveals many other uses of tunnels of which water supply, sewage disposal and mining are among the most obvious, along with tunnels to house power cables and tunnels for defensive or storage purposes.

The earliest tunnels were constructed for the purpose of water supply, and some middle-eastern settlements show evidence of such tunnels from as early as 1000 BC. Another early requirement for tunnelling arose within the mining process of extracting minerals from deep under the ground. While the aim in this case is to extract valuable material (which in other tunnels is useless spoil that needs efficient disposal), the requirement subsequently to support the resulting hole in the ground is clearly of paramount importance to the miners. Mining technology cross-fertilizes with other tunnelling technologies.

Many aspects of tunnelling can be illustrated by focusing our attention on transport tunnels. In the United Kingdom, canal tunnels were needed when the canal network was extended in the first part of the 18th century. The great age of railway construction in the late 18th and 19th centuries also required tunnels to overcome the problem of the limited gradients possible for conventional railways. The London Underground is the largest of several city rail subway networks, and remains in a developmental state. Road tunnels of various kinds form part of new routes, such as the M25, and of improved routes, such as the A55 in North Wales, as well as providing crossings of various major rivers.

In some circumstances, a tunnel will be the only solution to a problem. Attention then focuses rapidly on designing the particular tunnel. But a tunnel may be one possibility, and the merits of this solution then need debating against the alternatives – not least in terms of cost and environmental impact. One such debate in the UK has centred on the proposed extension to the M3 in the vicinity of Twyford Down where a cutting, although cheaper, would cause destruction of a valued environment in a way that a tunnel would not.

Any tunnel is the outcome of a technological process that has satisfied a perceived need. Its construction will have been proposed, argued, agreed and funded. A major project such as a new tube railway will have been the subject of a parliamentary approval process that has allowed scrutiny of the plans by those affected by its construction and use (e.g. people living above it). In the UK, the group requiring and funding the tunnel is known as the *client* (or promoter), and may be government, statutory or local authority or a private developer. A large client may have in-house expertise to undertake the project, but normally the client appoints a firm of *consulting engineers* with appropriate experience. In some instances projects may be sufficiently large or complex for the works to be split between two or more firms each with their own responsibilities. The consulting engineers undertake design of the works and generally supervise its construction through their *resident engineer*. A large project may be subdivided and let as a number of contracts each undertaken by a *contractor*, normally appointed following competitive tendering. The contractor's on-site *agent* is responsible for all aspects of the construction, including appointment of sub-contractors for any specialist work. Site-based project administration includes programming aided by computer-based critical path analysis, cost control, labour relations, safety and application of contractual and engineering skills in accordance with the contract requirements.

■ Technical aspects of tunnelling

A primary requirement of tunnelling technology is to ensure that tunnels are not only built exactly where required, but also that they subsequently maintain their structural shape and integrity. Both of these aspects need to be achieved within an environment where safety considerations are paramount. Other requirements include minimizing ground subsidence above the tunnel, restricting the ingress of water to manageable limits and ensuring that materials used in construction favour long-term maintenance. Tunnelling contractors can employ a range of techniques according to the different structural requirements of the completed tunnels and to suit methods of excavation appropriate to the anticipated ground conditions. Tunnels may be excavated by hand or with the aid of a variety of digging machines, or blasted, or driven using a tunnel-boring machine. They can also be constructed by cut-and-cover methods in which the square or rectangular structure is built in a prepared trench that is subsequently backfilled to ground level. This last technique was used for many of the early sub-surface railway lines of the London Underground. The immersed tube tunnel is a development of this method where prefabricated concrete sections are laid in a trench across a river bed as in the A55 River Conwy crossing in North Wales. Where depths of water are appropriate (as in the Strait of Gibraltar), outline consideration has also been given to submerged tube tunnels suspended in the water but anchored above sea-bed level rather than contained in a trench. In cross-section, tunnels may be circular, elliptical, horse-shoe, square or rectangular, depending on their function and method of construction. They may be unlined, or lined with wood, brick, steel plates, cast iron or precast concrete segments, or concrete cast, or sprayed *in situ*. A smooth lining may be advantageous in allowing water flow, but of less consequence in a railway tunnel. Not only have construction methods progressed from hand-digging to mechanized techniques, but ways have been developed of stabilizing surrounding ground during construction when conditions demand it.

Hard rock and soft ground present tunnellers with different challenges. In the former, an important requirement is to construct the tunnel reasonably quickly while in the latter this extends to maintaining the structural shape of the necessary lining. Ground conditions are notoriously variable across both location and depth and represent the main risk element in tunnelling. They can result in design changes during construction – as an example, the 2.4 km Blea Moor tunnel on the Settle–Carlisle railway line was driven through granite but had to be brick-lined throughout to prevent subsequent rock falls. Soft ground may consist of clay, sand, gravel, boulders, peat or any combination of these and frequently in association with water. Certain ground conditions may also contain gas. Long tunnels, such as London tube railways, may well encounter a range of conditions calling for different techniques to be employed at different locations.

Whatever the nature of a bored tunnel, it is clear that precision measurement in difficult conditions is necessary if bores approaching from different directions are to meet up. The opposing drives of the Channel tunnel service bore met to within 500 mm horizontally and 50 mm vertically after drives of 22 km from the UK side and 16 km from the French side. However, if we celebrate such an achievement with the Channel tunnel's laser technology, we should also spare an admiring thought for the builders of Hezekiah's 513 m water tunnel in Jerusalem whose digging teams celebrated their meeting with an inscription on the tunnel wall sometime in the eighth century BC.

■ A historical interlude

By the middle of the 19th century, hard rock tunnels were generally built using the mining method of drilling holes in the working face and filling these with gunpowder which was exploded, loosening a quantity of spoil which had then to be removed. Rates of progress were of the order of 10 m per month, with the tunnels often left unlined. In about 1860, the introduction of compressed-air drills and nitroglycerine enabled a significant improvement in the work rate to about 50 m per month. Nitroglycerine caused many fatalities until in 1867 Nobel perfected the manufacture of dynamite which combined the unstable nitroglycerine liquid with kieselguhr. The St Gotthard tunnel (1872) was an early major use of dynamite, as was Blea Moor tunnel. In this latter case, the dynamite had to be transported by road because the railways would not risk carrying it. Subsequent developments have been the tungsten carbide drill bit (*c.* 1950), which used materials technology to improve drilling rates, and the hydraulic drilling machine (*c.* 1980) which came into its own when the pneumatic drill seemed incapable of further development. Progress rates by this method are now reasonably expected to be of the order of 300 m per month. (Note that references in this paragraph to progress rates are intended to give a concept of order of magnitude. They clearly depend on tunnel diameter, exact nature of ground, etc.)

An alternative to blasting was developed in the 1950s in the form of tunnel-boring machines which use rotating tungsten-carbide tipped cutters to gouge out a circular bore through the rock. Such machines represent a drawing together of many aspects of engineering and science – not least to solve the considerable problem of removing the heat generated by such a concentrated source of power in an enclosed environment.

Soft-ground tunnelling in the UK developed largely from attempts in the early 19th century to tunnel under the Thames (necessary because bridges would prevent the access of ships to the docks). The engineer Marc Isambard Brunel found inspiration while watching a shipworm (*Teredo navalis*) drilling its way into a piece of timber. Engineers, as well as scientists, have their *eureka* moments! The lowly shipworm sparked off the invention of the shield – in effect a protective shell within which manual excavation of the workface could proceed. The shield prevented premature collapse of the surrounding ground, and the tunnel bore behind the shield could immediately be lined with bricks. The shield could then be moved forwards by jacking it away from the lining. Although Brunel utilized a shield, the name subsequently associated with shield tunnelling is Greathead, engineer of the Tower Subway built under the Thames in 1869. Greathead used a 2.3 m diameter circular shield and was also the first to use a tunnel lining of bolted prefabricated cast-iron segments (an idea borrowed from mine shafting). The technique of Greathead's shield, cast-iron bolted linings with any void (space between the lining and surrounding ground) subsequently grouted (filled up with a cement material), was subsequently used with great success on the construction of most of London's tube railways when that network grew rapidly between 1896 and 1907. The use of prefabricated tunnel-lining segments greatly speeds up progress when compared with linings built *in situ* with bricks and mortar. As well as being faster to install, the rigidity of the lining means that the shield can immediately be moved forwards with its jacks, rather than having to wait for the mortar to set. In current practice, shield-driven tunnels in good clay are usually provided with a mechanical means of excavation where the length justifies the cost of this approach. An interesting cross-fertilization of ideas is that soft-ground versions of tunnel-boring machines are now used for appropriate long drives in soft ground.

A particular problem of water seepage occurs when construction is carried out below the water table. The solution, first tried out in the Tower Subway, is to supply compressed air that raises the pressure inside the bore to a level that holds back the water (one atmosphere holding back about a 10 m head of water). This system means that workers have to pass through an air-lock and would be subject to problems like 'bends' akin to those of divers unless correct procedures were observed. The installation of a waterproof lining enables the air-locks to be removed on completion of construction.

The contractors who actually build tunnels are commercial firms operating in a competitive environment, and this leads them to look for ways of reducing costs. Faster construction and cheaper materials can both contribute to this, although the former must take account of worker safety and the latter of user safety. An industry as potentially disaster-prone as tunnelling is rightly rather conservative in its approach to new ideas. For example, in the early 1900s, McAlpines experimented with concrete tunnel linings and these were successfully used in sewers in the 1930s. When, in 1937, it was proposed to extend the London Underground Central Line, there was a general shortage of cast iron for civil use because of the re-armament programme. This provided the opportunity for the building of an experimental section from Redbridge to Newbury Park using precast concrete lining which was, nevertheless, a copy in concrete of the bolted-and-grouted cast-iron lining. This lining was found to be successful and cheaper that the cast iron and so the advocates of concrete were eventually able to overcome the conservatism inherent in the industry.

■ Case study: planning and designing the Victoria tube from Walthamstow to Victoria

☐ Introduction

The Victoria tube was open to public use between its northern terminus at Walthamstow Central and Victoria by March 1969, with an extension to Brixton opened in 1971 to give a total route length of 22 km. It was the first completely new tube built since 1907, and its opening satisfied a need originally identified in 1946 when the then Railway (London Plan) Committee recommended such a route for a tube railway. In providing a new north-east to south-west link across central London, the new line was expected to generate some new passenger revenue, but mainly provided relief for existing overcrowded routes and was expected to contribute to a reduction in surface traffic and an annual transfer of 26 million person-journeys from bus to tube. It was financed as a public service, as opposed to being set up as a project to generate revenue for shareholders. Figure 1 shows the geographical route of the line.

The desire to connect with the existing network of tube lines, and the unavailability in central London of ground-level space meant that, when this line was planned, there was no sensible alternative to a tube railway for most of its route. So problem-exploration focused on this one kind of solution. The project to design and construct the bored-tunnel tube from Walthamstow to Victoria had three main stages with dates roughly as follows: preliminary planning (1955–62), project planning (1961–62) and construction (March 1963 to March 1969). There is some overlap between the stages, and the rest of this article concentrates mainly on the preliminary planning stage. The client for the project was the Board of London Transport, which employed two firms of consulting engineers to act on its behalf. Various engineering contractors and sub-contractors were employed on construction work.

☐ Preliminary planning (1955–62)

The British Transport Commission was granted overall statutory powers to construct the line by an Act of Parliament in 1955. Such an Act was necessary before detailed planning work could start and it also empowered the compulsory purchase or temporary occupation of the land that would be needed for construction sites including shafts to various tunnelling faces. Parliament only allows construction work to start on such a project when it is satisfied with the main routing and constructional details, including the methods to be used. So the preliminary planning had to generate 'Parliamentary plans' that met the (then) Ministry of Transport regulations and specified enough detail for this decision to be made. A tunnel of this kind needs routing for both *line* (as it appears on a map) and *level* (effectively its depth below ground along its length). The two are related, because a desired line will have implications for level, especially when interchanges with existing tubes are required and when many other obstructions, such as large sewers, exist. For example, Figure 2 shows the line in the vicinity of King's Cross station.

In design terms, this means that certain desirable general characteristics of tube lines interact with a range of quite specific, localized constraints. The general route of this tube was not in question, but in several local instances the eventual route was chosen from several possible variations on the basis of revenue-earning traffic potential, quality of interchange facilities and cost of construction. Such decisions are made by the client using technical information provided by the consulting engineers and other information (such as passenger surveys). The consulting engineers will have provided detailed design drawings and information on costings and schedules for the various alternatives, as well as any foreseen requirements such as availability of specialist plant. Such decisions represent a classical example of the need to adopt a multi-dimensional approach to solving a technological problem.

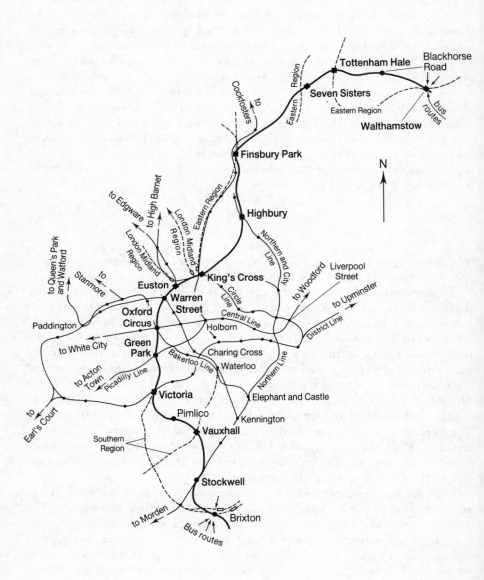

Figure 1　Victoria Line and existing routes.

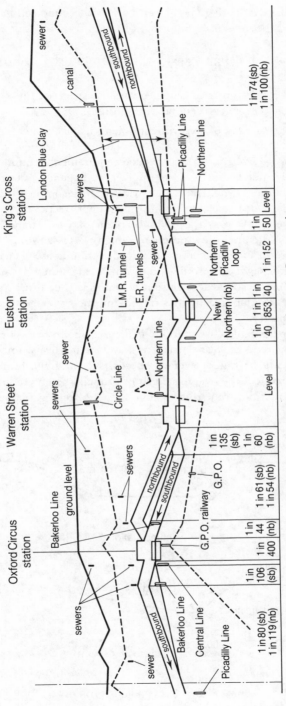

Figure 2 The line in the vicinity of King's Cross station.

As well as data gathered on local traffic and ground conditions, the design of this new tube line used information from various sources as follows:

(1) Accumulated design expertise relating to the construction of tube railways.

(2) General research carried out into various aspects of line capacity (the rate at which passengers can be transported).

(3) The construction of experimental lengths of tunnel.

Accumulated design expertise   The particular requirements of, for example, tunnel alignments at interchange stations interact with general desirable features of tube running lines. Gradients facing trains are recommended as not steeper than 1 in 60 for a distance of up to 400 m and not steeper than 1 in 100 for any distance above this. From the drainage point of view tunnels should not be level, but should contain low points to which water can drain and be pumped away as necessary. From the point of view of economic use of electrical energy, a hump station or sawtooth profile is recommended (Figure 3). As well as saving energy, this improves the acceleration of trains as they leave stations and their deceleration as they approach, thus improving their average speed and assisting with line capacity.

Another desirable characteristic is for the line to have curves that are not too tight. This reduces wear on the track and train wheels, improves passenger comfort, and allows higher average speeds. Keeping the route as straight as possible also reduces the overall length of tunnel, thereby reducing the overall cost of construction, but this option is not often available due to other

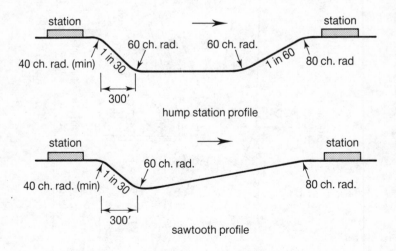

Figure 3   Gradient profiles.

requirements. Many of the older tube lines had very tight curves, particularly where they followed the street layout above. This was done in an effort to limit possible compensation paid to owners of properties that might be affected by the proximity of the tunnels. By the time the Victoria Line was being constructed, experience had shown that the presence of tubes under properties in London could be tolerated. However, route alignment had to take account of a new generation of tall buildings having particularly deep piled foundations.

Line capacity research Maximized safe passenger-carrying capacity is clearly a desirable feature for a tube line of this kind. At the time of this project, some aspects of train and passenger movements could be represented by appropriate models. For example, a mathematical model shows that line capacity depends mainly on the frequency of the trains, which in turn depends on length of station stop times, the distance between stations, the signalling system, and the speed, acceleration and braking characteristics of the trains. For a new line, all of these variables (with the possible exception of distance between stations) are subject to design decisions and each can be further itemized: for example station stop time at a busy station depends on access and egress connections at platform level, carriage layout, number of doors, etc., all of which contribute to the free flow of passengers and reduce bunching on the platforms. The signalling system and dynamic characteristics of the trains determine the minimum safe headway (the time between two trains) and speed-control signalling can compensate to an extent for longer station stops by allowing a following train to approach one already in a station, rather than having to stop at a greater distance and then re-start.

The mathematical model in this case postulated a 30-second station stop during which 900 people could pass through the doors of an eight-car tube train, assuming uniform passenger distribution along its and the platform's length and a regular boarding/alighting pattern. Such assumptions immediately reveal the limitations of such a model even when applied to one train stopping in one station. But this single-station model assumes maximized passenger flow along the platform by having entry and exit at opposite ends of the platform. A conflicting requirement becomes apparent when it is recognized that, to avoid bunching of passengers, entries and exits at adjacent busy stations should not be at equivalent positions relative to the train (i.e. one at each end). So the model has its internal conflicting requirements, as well as being faced eventually with the realities of interconnections with existing tubes and sensible routes to surface exits. It is worth pointing out that the mathematical model was itself informed by empirical results from practical experiment — for example, the rate of flow of people along constricted passageways was investigated both *in situ* and with the help of numbers of schoolchildren who took part in controlled investigations into movement along passageways of various widths. The movement of passengers on escalators and staircases had also been the subject of practical experiment.

Experimental lengths of tunnel   London tube railways are built, where possible, in the clay layer (see Figure 2). Most existing tubes (and their recent extensions) had used 3.66 m diameter bolted cast-iron linings. When the Victoria line was being planned, there had been some experimentation with cast iron and concrete linings that needed neither bolting nor grouting. Such linings would be cheaper and faster to install than conventional ones, so were attractive. The rate of tunnelling of the bore in clay can be so rapid that the rate of assembly of the lining can be the factor limiting progress. Theoretical work had been done on the stresses set up in tunnel linings as the ground settled. In 1955, a section of disused tube constructed of bolted cast-iron segments was used for an experiment in which strain gauges were mounted and the lining bolts subsequently loosened and removed. When the removal of the bolts produced negligible movement of the lining segments, this practical experiment confirmed the prediction from stress-analysis that the tunnel lining itself was unlikely to distort greatly with time. The way was open for consideration of non-bolted linings – another step that would speed up construction. It is the consistent circularity of the bore produced by a shield in clay that allows the tunnel lining to be placed directly against the exposed clay with no need for grouting between the lining and the clay. But the lining must be flexible enough to take up slight movements as the ground re-settles. The paramount need for a tube tunnel is that the lining shall not collapse or allow ingress of water (beyond any small seepage that can be pumped out), so any departure from conventional practice needs thorough preparation, part of which was, in this case, to build experimental lengths of tunnel.

It was the responsibility of the two firms of consulting engineers to specify the nature of the lining, and each firm did this partly on the basis of the experience of building these experimental lengths, which had also drawn on recent experience gained with other tunnels. The internal diameter of the running tunnels was selected as 3.81 m with a concrete lining thickness of 150 mm. Various diameters were considered, up to a maximum of 5.2 m which would have enabled larger rolling-stock to be used. However, this diameter would have greatly increased the cost of tunnelling. 3.81 m was chosen (as opposed to the standard 3.66 m) because, using standard-dimensioned tube rolling-stock, the air resistance on a moving train would be about the same as that in the open air, thus minimizing energy requirement in this respect. Each ring of lining in this case had 11 main segments (Figure 4), with design variations to accommodate the track bed, cable supports, etc.

Parliamentary plans

Once the route was agreed and the engineering investigation completed, the Parliamentary plans were drawn up based on ordnance sheets corrected for detail by reference to an aerial survey of the route. The route plans showed the proposed line of the tube within limits of deviation of total width 90 m. 3300 properties were identified at this stage as being threatened with some kind of interference (tube passing below, compulsory purchase to make way for surface work, etc.), but only 26 petitions against the proposals were received and these were all dealt with before the Bill came before the House of Commons.

☐ **Project planning (1961–62)**

Detailed planning was stepped up in 1961 on completion of the experimental tunnel lengths. At this stage, final decisions had to be made that drew on the earlier investigative and design work, but that also drew on more detailed geological investigations. A necessary feature of the project was that preliminary *programming* (i.e. initial decisions about what to do during construction) had to be done before construction was finally authorized. So, for example, the shield drives (the driving of the main tunnels) were planned to take place from working sites identified within the Parliamentary plans. These working sites were selected not only to give suitable access to the tunnels etc., but also to minimize disruption at ground level. An initial requirement for 30 shields was envisaged, allowing work to proceed simultaneously in this many locations. The programming needed to ensure that no one job would delay final completion, and also needed to allow for certain sequences of jobs, especially

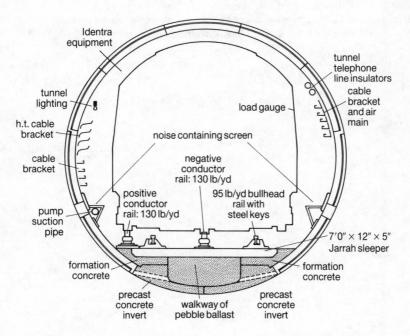

Figure 4 Running tunnel: 3.81 m precast concrete lining.

where existing services had to be maintained where other lines or passenger subways had to be diverted.

Before final authorization, seven preliminary programmes for the project were considered and the one adopted envisaged a three-stage opening of the line to allow sensible phasing of some aspects of the work. Tunnelling contractors were introduced to this programme and advised as to how the work was to be divided before they were asked to tender for particular contracts. This enabled them to assess the demands on their estimating departments (in preparation of tenders) and on their more general resources (in carrying out any contracts awarded to them). Civil engineering construction was divided into three contracts for preliminary work and 20 main contracts for construction.

■ Acknowledgements

Material for this article has been obtained from:

Follenfant, H.G. *et al.* (1969) *The Victoria Line*, Paper No. 7270 S, Institution of Civil Engineers, London.
Turner, F.S.P. (1959) *Preliminary Planning for a New Tube Railway across London*, Paper No. 6343, Institution of Civil Engineers, London.
West, G. (1985) *Technical innovation and the rise of the modern tunnelling industry*. Unpublished PhD thesis, The Open University.

3.6

The United States Air Traffic Control System: Increasing Reliability in the Midst of Rapid Growth

T.R. La Porte

[This edited account of the USA air traffic control system (USATS) traces the development of an organization that manages a growing volume and complex mix of traffic with increasing scope, safety and reliability. We include it to illustrate the features of technologies where considerations of 'system maintenance' are important. Large technical systems (LTS) must take account of the necessities of those who operate the systems, and this article explores some of these necessities.]

■ Introduction

United States air traffic system (USATS) providing both air navigation and traffic separation became a nationwide governmental service in 1936 after two decades of expanding private and public activity. Within 50 years, this system has grown into an extraordinary matrix of 600 airports and 300 000 miles of airways in continuous flux and motion as millions of people and mountains of freight (and air mail) are shepherded throughout the US. It has been a remarkable development of a very large-scale, publicly owned technical system [. . .] After a brief review of the dimensions of the USATS, we turn to properties [that are important] for more general understanding of large-scale technical systems, and go into more detail in describing the extraordinary development of the USATS.

The initial stimulus was transporting mail by air. Both early airmail and airways services were managed by the US Post Office Department until 1925, when private contractors took over the mail services. Air mail flights had expanded from the first regional (daylight) links in 1918 between New York and Washington, DC, to reach across the continent with the development of night-flying navigation aids. Rotating beacon lights set up every 10 miles guided low-flying pilots over 2000 miles of lighted airways between New York; Dayton, Ohio; Chicago; Cheyenne, Wyoming; to San Francisco (with a spur to Los Angeles).

[. . .]

Our attention centers on 'United States Air Traffic System' (USATS). It is a web of technologies and institutional relationships linking the components of the larger US air transportation system through continuous co-ordination of aircraft. The system's primary institutional embodiment is the Federal Aviation Administration (the FAA) and its predecessor agencies. Secondary notice is taken of the air carriers and other 'users' of the system.

The USATS, unlike EUROCONTROL, its younger and much smaller brother in Western Europe, is predominantly funded by resources from the general tax fund. Conceptions of economic development do not adequately explain USATS development.

[. . .]

The time frame of this review is limited, beginning with the early days of the system in the 1940s and ending in 1980, just before its third major institutional crisis – the tumultuous strike of the Professional Air Traffic Controllers Organization (PATCO). [Table 1 shows the changes in elements of the system over that period.] This strike, its aftermath in operational travail, and the recent problems of the FAA (brought on by a combination of the deregulation of air transport and a controller cadre working continually at or near full capacity) are fascinating in their own right. Understanding this crisis, however, requires a good bit more than the story discussed below.

■ An integrating frame

A major step toward integrating technical and social science perspectives can be taken by conceiving of technical systems as social organization. In this view, the technical design and operational imperatives become guides to operator and managerial behavior. From a social science (or public policy) view, unless a technology becomes widely spread (or is likely to become so) it is a trivial activity. Widespread distribution or deployment of a technology necessarily requires some form of large-scale social organization. It may be decentralized as in the manufacture and distribution of personal computers. It may be physically and organizationally widespread and highly integrated like the distribution of electrical energy through large regional, national or even multinational grids.

In this view, the techno-organization animates or gives social expression to technical possibilities. This perspective challenges us to examine the properties of technical designs and engineering systems in terms of their organizational

Table 1 Elements of air traffic system and changes of scale: 1940–1985.

	1940	1980	[1986]
Airports:			
(paved, lighted)	776	5 830	6 720
Aircraft:			
Prop	—	238 160	246 540
Jet	—	5 869	8 174
Air travel (in 1000 hours):			
Domestic air carriers	710	6 250	7 360
General aviation	3 200	41 000	34 063
(Revenue passenger miles in millions)	1 050	200 000	270 100
Air traffic control:			
Airway miles (1000s)	32k	296k	325k
Navigational aids (all types)	340	2 090	2 261
Landing aids (all types)	—	988	1 166
Facilities (terminal/route)	11	527	525
FAA employees (1000s)	5k	55k	47k
Aircraft handled per yr (in 1000s)	(1945)		
Air carriers	2 610	23 600	26 373
Air taxi	—	7 230	11 794
General aviation	410	36 720	30 523
Military	2 610	5 990	6 328
Total	5 630	73 540	75 020
Safety record:			
Air carriers (domestic operators)			
Accidents per 100k hrs	4.2	0.22	0.22
Fatal accidents	3	0	4
Fatalities	45	0	197
Fatal accidents per 100k hrs	0.42	0	0.05
General aviation			
Accidents per 100k hrs	108.4	9.2	8.6
Fatal accidents	232	618	490
Fatalities	359	1 239	937
Fatal accidents per 100k hrs	7.2	1.7	1.53

requirements and imperatives. It leads us to explore the relationship between the designers' views of operational necessities and the implications of implemented designs for the behavior of operators who man the system.

[. . .]

Functionally, the USATS is a complex 'sub-system' of the larger 'whole system' of the US air transportation industry. It is a lesser included, crucial element, in air transport operations. It is also much less fully integrated with its system neighbors than the elements of other systems.

Airplanes and pilots can operate with more autonomy than trains, telephone services and electrical power systems. The connective networks are much less dominated by physical objects – rails, wires and power grids.

Finally, an air traffic (sub)system is largely a mental rather than a physical construct. It has no visible, concrete supporting connectors. The system must be 'seen in the head', a mental construct recognized by thousands of people (controllers, pilots, facilities managers) in order for 'it' to be operative. US Air Traffic Control (ATC), the operator/controller of the USATS, is the arbiter of the mental maps and procedural agreements guiding the behavior of its members. These are quite detailed, with many critical aspects, and must be known and followed by many, many users in order for the system to work highly effectively and reliably. This aspect is much less evident for telephones, electrical circuits, or railroad systems.

[. . .] It is important to keep these characteristics in mind as we describe some of the salient aspects of USATS development.

■ The development of USATS: external and internal guiding dynamics

The USATS has had an almost unbroken path of vigorous expansion. Such a pattern requires, at least, a high degree of agreement on system purposes and functions. Throughout its history, the USATS has been the object of an extraordinarily high degree of consensus about its mission. All of the major actors within and outside of the system have agreed that:

- Flying is intrinsically valued and air travel produces a major social benefit.
- All those who wish (and can afford) to fly should have the technical and operational means to do so.
- Due to increased demand for flight, increased technical capacities for aircraft, airports and co-ordination of aircraft aloft are required. It is the responsibility of the Federal Government to assist this development.

There has been an underlying political agreement that access to air travel via either private means or commercial carriers is very nearly a public right. (This has only recently been questioned.)

The result of this consensus has been a readiness, if not always an ability, to respond favorably to proposals for increased resources for development. Indeed, during the time of our interest, the US Congress had never reduced the amount of money requested by the FAA in support of their air traffic control function. Favorable treatment depended on the degree to which needs could be established and programs justified on the basis of meeting operational criteria. These criteria set the framework for the logic of development, and shaped the character and intensity of energies propelling organizational growth.

External demands from the host society have been constant, if potentially contradictory. The public (and especially its Congressional leaders) demands a system which:

- is always safe
- carries anyone, anywhere, anytime (and is always safe)
- enables private carriers to make a reasonable profit (while always being safe)
- requires only modest co-ordination expenses of carriers, and the flying public. (Secondarily, keep costs of governmental administration moderate in terms of the level of safety and ease of traffic movement provided.)

From the earliest days of air travel in the US, there has been a strong emphasis on reducing the risk of operating an inherently hazardous technology. The economic success of air travel depends, in part, on the public's perception that using the service 'can be habit-forming', i.e. one can do it time and again and survive.

[. . .]

The twin pressures from the travelling public and elites for extraordinarily reliable and safe performance resulted in a system — one of several large technical systems in the US — that has attempted to achieve failure-free operations. That is, the goal of failure-free performance is a central objective of everyone in the system. This drive to achieve very high levels of operational reliability and the demonstrated effectiveness in nearly reaching these goals year after year qualifies the system as a 'high reliability' organization. It is a quality that has had an overwhelming impact on the character and shape of the system's evolution.

Technical systems, then, are initially shaped by the operating requirements and social properties of technical operations that are inherent in its technical design. In the operation of the air traffic system, these imperatives were (and remain):

- *The technical and operational imperatives* to provide accurate, unequivocal information about location and intention of every aircraft; procedures which eliminate or drastically reduce the likelihood of disoriented aircraft or unexpected convergence of aircraft aloft, and assure timely guidance information to aircraft operators so that no aircraft 'loses separation' from another or has a near collision or, most especially, a mid-air collision. The operative goal is to avoid 'loss of separation', i.e. to allow two aircraft to come closer than five miles apart (and 1000 ft in vertical separation). This is an absolute criterion for controller performance. If a controller suffers a moderate loss of separation between two aircraft he/she is working *three times during their whole career*, they are discharged.

The technical and managerial imperatives to expand an integrated network throughout the nation and strive for optimum internal activity, interaction, and density of flow. The result was/is efforts to 'pack the system', specifically, to press for headway between aircraft just above legal separation limits − now five miles at altitude, and three miles near airports under visual flight rules (VFR).

This combination of 'imperatives' leads to a fundamental and abiding tension between safety and reliability on the one hand, and efficiency, on the other. In operational terms, tensions are between those who directly benefit from perceptions of safe systems − commercial pilots, air traffic controllers, Congress (and passengers) − and those who must pay for it − air carriers, general aviation pilots, and the administering agency, its political/budgetary overseers. Users press for the resources and regulations necessary for totally safe commercial flying conditions; payees worry that the technical and regulatory safety and capacity requirements are more costly and constraining than necessary to keep air traffic moving economically and safely. [. . .] This is a rich stew of advocates and watchdogs. It is fruitful ground for conflict over means and has the potential for exploitation. Much of the development story of US Air Traffic System reflects such dynamics.

■ The development of USATS: growth and consolidation

The USATS' maturation has been characterized by strong technical advocacy, institutional turbulence, extraordinary growth and astonishing reliability. [. . .] Operational requirements consist of maintaining a cadre of dedicated air controllers and airway facilities employees who give social animation to the technical systems of communications, electronics and procedures. Technical planning and development requirements call for advanced engineering, solutions to demanding (and interesting) technical problems and the deployment of costly new systems likely to change the working conditions of the operator cadre (and alter their relationships with pilots).

Early FAA leadership was in full accord with both Congressional and industrial leaders: increase the use of air transport (rail transport was the implicit comparison). There was a vigorous program of airport construction and improvement, and, in the pre-war late 1930s, a sense of urgency and then action to promote the growth of aviation infrastructure in preparation for hostilities. Early technical developments of air-to-ground communication, low frequency radio ranges and standardization of procedures for flying by instrument flying rules (IFR) had improved the capacity to identify and locate precisely the flight path of an aircraft. Controllers were trained to use co-ordination procedures and 'flight strips', manually enter a paper strip for each aircraft aloft, then track the aircraft across airways, routing it in place in the sequence of other aircraft before and after it. These capacities and procedures improved service and allowed effective co-ordination among aircraft separated by a minimum of 10 minutes or 10 miles headway separation. The system – in the midst of its first major technological phase – was established and 'in equilibrium' just prior to World War II.

The war brought substantial increases in traffic, technical developments and institutional challenges that set the stage for the FAA's first crisis. The character of the first crisis typifies subsequent problems and developmental dynamics. FAA and military responses to national defense requirements resulted in rapid expansion of communication service networks within the US, the deployment of FAA personnel to operate airport air traffic control towers to facilitate defense activities, and the establishment of provisional rules of air navigation. Military needs overwhelmed all others and the FAA functioned in large part as a civilian adjunct to military aviation and defense requirements.

During the war, military aviation developed new air navigation and air traffic technologies complementary to those of the civil aviation system. Military systems advanced beyond those employed for civil aviation, especially with the development of radar and its capability of 'seeing' aircraft many miles from an airfield. Military commanders became *de facto* managers over many in the civilian controller cadre. In 1946, immediately after the war, there was a rash of activity attempting to reorient the management of US air traffic system for peacetime conditions. As the system had grown, it had become dispersed and its management structure ambiguous. It was time to re-assert civil control of air traffic management.

The Department of Commerce was authorized to take over the operation of military air navigation facilities overseas. Scattered administrative and training units were consolidated in Oklahoma City, where all the FAA schools were to be centered. Joint research and development policies were established to assure continued technical development and the application of military technologies to civil air uses. Common civil–military instrument flight rules (IFR) were officially issued. The President established the Air Coordinating Committee by Executive Order with the responsibility for co-ordinating national aviation policy. The International Civil Aviation Organization (ICAO), the authoritative international standard-setting body, assembled representatives of 60 foreign states for a demonstration of US air navigation and traffic control equipment and techniques at the FAA's Evaluation Center in Indianapolis, Indiana. This move was influential in ICAO's later decision to recommend acceptance of the US systems and techniques as international standards.

[. . .]

Before World War II, airways were not crowded; the problems of safety were not yet closely related to the real likelihood of mid-air collisions. However, the rapid growth of aviation activities, the blossoming of military facilities and activities during the war years, and the general reluctance to raise post-war types of administrative matters until the war was over resulted in a general sense that the system could become inchoate and disorganized as demobilization got underway.

For some technical 'systems', e.g. the automobile or aircraft production, a 'disorganized' sector means freedom to compete, possibly to prosper. Monopoly or finely grained co-ordination, the intent of the 1946 developments, is not preferred by those who stand to gain from competition. In the case of USATS, we see another tendency: the drive to reduce sources of ambiguity or conflict that might be the root of operational surprise. It is a tendency likely to be shared by all technical systems that have a relatively high level of perceived hazard.

Technical developments also serve to reduce operational surprise. In addition to institutional coalescence, 1946 was the year in which perhaps the single most important technical advance in air traffic control was introduced — the radar-equipped control tower for civilian flying. This technology was first installed at Indianapolis Airport. (It was a modification and up-grade of radar developed by the armed forces.)

This development signalled the end of the first major technical phase of US air traffic system development. The predominant co-ordination technique had been a manual/voice reporting system of 'flight strips and shrimp boats' (small cutouts moved about a navigation map tracking the location of an aircraft as reported by the pilot). The manual/voice system would be supplanted either by a combination of radar, improved high-frequency navigation aids (VOR) and instrument landing systems (ILS) to improve pilot control during landings in foul weather or by 'ground controlled approach' (GCA) in which the aircraft was 'talked in' by operators scanning the plane's location and glide slope on specially designed radar. Radar would greatly improve the capacity of ground personnel to identify and assist aircraft aloft. As importantly for the development of the airways system, the *omni-directional* VOR capability exploded the number of courses available for navigation. [. . .]

In a sense, the original system had nearly filled up. With the generous margins for error necessary in the manual/voice reporting-based system, peak time air traffic near the most used airports was approaching full capacity. Increased system capacity was required. Radar, the new instrumentation and added radio and telephone communications between control centers provided it. This enabled controllers to increase substantially the number of planes that could be worked safely.

New technologies made the controller task less problematic when handling individual aircraft but more complex when dealing with up to 12 to

15 aircraft simultaneously. It also raised the question of how to deal with the situation if the newer, more sophisticated, more vulnerable technology failed. Would the controllers be blinded? How could they re-establish their picture of where everyone was? As radar was introduced, the original system was not replaced. Rather this non-electronic, 'cannot break' system is still manned and exercised, operating in parallel with newer systems, 'on call' as a continuously available backup.

Radar gave controllers an independent source of information on the location and disposition of aircraft. The first relatively primitive, sweep radar was augmented by a series of technical changes that systematically reduced the controller's dependence on aircraft captains for flight information. It thereby increased air traffic safety and reduced pilot autonomy.

As radar was deployed to airports and the air route traffic control centers (ARTCC) that monitored the airways between airports, the whole system could handle more aircraft simultaneously. The radar surveillance system was complete: the skies rapidly became more crowded. [The number of employees, financial resources and the actual use of air traffic control services do not show parallel growth. They are disjunct for three periods and point to times of strain and change. Each of these periods is discussed below.]

The Korean War produced the first period of strain. Air activities increased over 100% from 1950 to 1956, while the FAA's budget and manpower levels declined significantly. The FAA was again part of a war effort and controllers, most of whom had been in World War II, buckled down and kept the system together. It was a time in which technical changes and increased traffic flows would significantly complicate air traffic management tasks. In 1951, the number of air passenger miles first exceeded rail-sleeping car passenger miles (10.7 to 10.2 million). In 1953, airplane speeds could average over 200 miles per hour. In 1956, the first large jet liners carrying over 100 people were certified. In effect, the stakes involved in commercial aviation had doubled: twice as many people could travel twice as fast and twice as high as in the early days. This was tragically demonstrated high in the western skies in June 1956.

Two commercial airliners flying in the clear, deviated from their normal route to show their passengers glimpses of the Grand Canyon. They collided, killing 128 people. A Congressional investigation resulted and a series of restrictive measures were imposed to control the movement of aircraft at high altitudes. A continental airspace control service was instituted by the FAA requiring all aircraft in IFR conditions (in clouds) above 24 000 ft to be under positive ATC control. (Submission to this service was optional in clear air.) In 1958, a series of three more tragic airline accidents in the New York/New Jersey area triggered a Presidential investigation and resulted in recommendations for positive air traffic control on the main airways across the US. For all aircraft flying between 17 000 and 35 000 ft (this included all jet traffic), IFR rules conducted under prior clearance would apply. Visual flight rules (VFR) were rejected in these airways regardless of weather. These changes combined to increase the number of aircraft required to use ATC services and lowered the altitude above which aircraft control was required. The result was a sharp increase in controller workloads, and stimulated a need for more controllers.

At the same time, a battle was brewing between civil and military aviation circles. Research and development on more powerful navigation aids was going apace by both the FAA and the armed forces. In the early 1950s, the FAA had begun to deploy a much improved Very High Frequency radio beacon (VOR) that greatly improved the accuracy of determining and following specific directional headings and allowed for a considerably more complex airways system. It also had a distance measuring estimating (DME) capability which gave an indication of the aircraft's distance in miles from the radio beacon. Military development groups were developing a different system, the Tactical Air Communication and Navigation system (TACAN), with similar features, but employed different principles and was more robust for the varied types of operating environments they expected, especially aircraft carrier operations.

In 1947, Congress had directed that future technical developments should strive for a single integrated system. Views were fixed and for eight years (1948–56) progress on determining the single system stalled. Efficiency flagged in the face of technical aggressiveness and stubborn operational argument. In effect, redundancy was enhanced despite the best efforts of Congress and the White House.

Problems with the civil–military relations were not limited to technical rivalry. The Air Force and Navy still carried out a number of air traffic control functions. On the grounds of maintaining capacity for use in wartime, they wished to keep them. Some way of co-ordinating and rationalizing the use of facilities and integrating military and civil air traffic functions was needed so they would be compatible with national defence needs.

[A new FAA Administrator, Air Force General Elwood] Quesada had the job of consolidating civil aviation services and conducting Project Friendship, i.e. to negotiate what military facilities, practices and operations would be transferred to the FAA. This was completed in two years with the transfer of over 2000 military air traffic control facilities in over 300 global locations to the FAA. The lineaments of the present system were in place.

By the early 1960s, the jet age was maturing. A number of jet aircraft had been certified that carried well over 100 persons. Jet speeds were increasing. Air passenger transport had forged well ahead of both the railways (for long haul domestic travel) and ships (for the Atlantic crossing). In addition to much higher aircraft speeds and flying altitudes, further technical and system enhancements were made.

In 1958 and 1959, the FAA had instituted Continental Control Areas (above 24 000 ft) and several Positive Control Routes (between flight levels of 17 000 and 35 000 ft) in which aircraft were mandated to be under instrument flight rules (IFR), have operative radar and radio communications, and place themselves under ATC direction. By 1961, this system was replaced by a national system providing routing direction and radar advisories along three tiers of airways: lower level from 1200 to 14 500 ft, intermediate airways from 14 500 to 24 000 ft and high altitude jet ways above 24 000 ft. At about the same time, computers were beginning to be used for aircraft accounting tasks.

This three-tiered airways system enabled the FAA to continue serving a rapidly growing aviation industry within a traffic system which had become increasingly dense and tightly coupled. It also further complicated air traffic control operations, and required a parallel division of labor within ATC centers. Total FAA employment had increased to about 30 000. The air traffic system had become a full fledged bureaucracy of sizable proportion.

The FAA established Associate Administrators for Administration, Programs and Development. The airspace system programs included the Air Traffic Service, Flight Standards Service, Systems Maintenance Service and Airports Service providing guidance to seven US FAA regions. Then, partly in response to the increased co-ordination needs, the FAA reversed a 15 year policy and gave the Washington office direct supervision over programs in the field. This immediately preceded several years of increasing operational and administrative complexity when air route traffic control centers (ARTCC) were being upgraded technically and, as a consequence of better radar and communication capabilities, were consolidated into fewer, more widely ranging ARTCCs.

The mid-1960s brought the second [. . .] period of strain. Between 1963 and 1967, there was some 65% overall growth in the amount of ATC traffic. Resources, however, did not follow the same pattern. The FAA's resources and manning levels *dipped* some 10%. The US had become embroiled in Vietnam and war costs were soaring. President Johnson was attempting policies of both 'guns and butter' and many Federal agencies faced increasing demands for services with stable budgets. While personnel resource levels hit a plateau, work loads increased steadily, due about equally to growth in both commercial and general aviation users.

At the same time, technical, procedural and administrative changes were 'rationalizing the system'. By 1964, three-tiered airways gave way to the present two, and DME (distance measuring equipment) was mandated for all civil aircraft flying above 24 000 ft. Solid-state, real-time computers were introduced throughout the system. As a result, ATC operations were modestly simplified. Advanced radar systems increased the accuracy of aircraft position images. Computer-generated displays of aircraft identification and position enabled controllers to increase the number of aircraft they could handle simultaneously from 12–15 to 20–25. By 1965, the Continental (positive) Control Areas were expanded to cover the whole US. These technical and procedural improvements increased individual controller effectiveness.

Communications and administration were also improved. By 1968, the FAA had put in place a nationwide telephone and telex system connecting the central office with the most active airports, ATC regional offices and area managers. Daily conference telephone calls became a standard feature of management co-ordination. In addition, the FAA and Air Force increased their co-ordination and eliminated overlaps. ATC played a larger role in defense interception work and Continental Defense Command activities. Key FAA programs were centralized under the Administrator, while several critical functions that varied from region to region, such as the designation of controlled air space in terminal areas, were de-centralized.

Demand, however, grew faster than the system's capacity to handle the volume with ease. The ATC system geared up to handle increased demand. It extended the amount of controlled airspace and improved more airports to enable them to receive ATC co-ordinated aircraft. Yet budgets and manpower allocations remained relatively constant. The few modest increases were used for capital and computer purchasing programs. The system became more densely packed, the margins for error declined, and working conditions worsened.

This situation drove controllers to consider organizing to secure relief from increasingly demanding, fatiguing and harrowing work conditions. FAA management was unsympathetic. The controller cadres were expected to perform in the face of adversity. They were then and still are part of a 'can-do' organization. In many respects, they had a number of the characteristics of a quasi-military management culture. And they endured these conditions for some five years after the onset of the 'stable state'. In 1968, after considerable internal debate, the Professional Air Traffic Controllers Organization (PATCO) was formed with a membership of 5000 in the first year. (It was to grow to over 15 000 by 1980.) [. . .]

Shortly after the formation of PATCO, the ATC system experienced its first instance of extreme airport congestion when the New York area airports had a day in which almost 2000 aircraft were significantly delayed in taking off or landing. For the first time, the FAA was put into a position of having to restrict the use of certain airports. This was the initial break in the FAA's long-standing public policy of serving any pilot who sought assistance at the time he/she requested it. The agency was edging into a position of having to ration its service − a process it still has a difficult time carrying out.

During the 1960s, the goals of service to all in a climate of extraordinary safety led to a series of incremental improvements in new technical systems, changes in procedures and air use restrictions, and operating rules that brought considerably more airspace under direct FAA control, e.g. through lowering of Positive Control Area altitudes from 24 000 to 18 000 ft, and raised the spector of perhaps having to assign priorities to different classes of aviation. This, in turn, raised the question of the optimum relation between serving commercial, highly professional air crews and companies contrasted with the much more numerous, generally less well trained and equipped, though increasingly well organized association of general aviators.

There was and is the general recognition that safety problems arose primarily from pilots who were less skilled and/or were not under direct control of the ATC aloft. This was the source of the unidentified, surprise aircraft suddenly appearing on the radar scope or inadvertently entering restricted airspace and tangling with a commercial carrier. These were almost inevitably

General Aviation pilots, i.e. private and business employed pilots flying unscheduled, irregular flights. [. . .]

There has been a steady trend – continuing to the present – toward expanding the positive airspace under mandatory ATC control and increasing the instrument flying skills and navigation equipment requirements, e.g. multiple radios and radar transponders, in order to obtain ATC services. In the interest of overall efficiency and safety, users of the system have been required to increase their skill levels, technical and equipment capabilities and procedural and operational complexity. These changes have squeezed out the General Aviation aviator who has neither the time nor the money to keep highly skilled and to purchase and maintain costly on-board electronic equipment necessary to qualify for ATC service.

The benefits, stakes and costs of reliable, effective air transport were steadily growing. Thus far, however, sharp trade-offs in service had not been necessary. Vigorous activity was continually required to stay ahead of the demands of increased traffic. Higher skills, more information and tighter co-ordination processes were also necessary to handle increased system complexity and density. Computer-based data links and inflight following and updating of aircraft progress were improved. And more finely integrated landing and navigation systems were introduced.

In the late 1960s and early 1970s, the FAA paid greater attention to the improvement of ATC controller training and retention. The agency expanded its national ATC training facility. Measures were taken to improve controller work situations. These changes came at a time when PATCO first tested its strength by initiating a three day, small scale, relatively ineffective work stoppage or 'sick-out' in June 1969. The 'sick-out' was followed by the organization's first formally called strike in mid-1970. Some 3000 (of some 16 000) controllers, mostly at the key ARTCCs, walked out for nearly three weeks. Airline schedules were severely disrupted. The issue, as in the earlier 'sick-out', had to do with working conditions, pay and benefits. Having made its point, PATCO called off the stoppage during the court-ordered showcase hearing.

Another technical/systems development advance, Central Flow Control (CFC), was quietly introduced at FAA headquarters in 1970. CFC has been critical to the increased co-ordination of the sprawling ATC system. This facility took over some of the responsibilities of controlling the flow of traffic from the 21 ARTCC centers throughout the US. Linked by telephone and teletypewriters, the facility was able to determine the overall capabilities of the system on a daily basis and issue instructions for restricted air traffic flows into areas that fell below expected capacity. (CFC became immensely important in the FAA's response to the near national emergency precipitated by the firing of 11 400 PATCO controllers in 1981.)

The third period of strain occurred in the latter half of the 1970s. General Aviation levels exploded. While commercial carriers were more or less constant in their hours of flight time, jumbo jets were introduced. The passenger carrying capacity for commercial flights almost doubled, up to 200–250 per flight and flying speeds rose dramatically. Once again, the stakes involved with safe flight escalated.

The system approached another period of expected saturation. Brisk planning went on in anticipation of changes in the 1980s. A National Airspace Plan was devised which was intended to provide the radar and computer technologies to 'tighten the system' even more, packing more aircraft into the airspace, with more finely co-ordinated traffic control in metropolitan areas hosting an increasing number of airports with enhanced landing capacities.

Air traffic levels continued to increase dramatically. During the same period, FAA financial resources declined in constant dollars. Personnel levels declined as well. The stage was being set for a conflict between controllers and management. This time a robust union was in place.

◼ Conclusions: properties of networked technical systems

From this review of USATS development, what can be learned about the developments of networked large-scale technical systems? [They] are:

- Tightly coupled technically, with complex 'imperative' organization and management prompted by operating requirements designed into the system. [. . .] (This is a kind of soft technical determinism: either do it my way or it won't work and do good things for you.)

- Prone to the operational temptations of network systems, i.e. drive to achieve maximum coverage of infrastructure, and maximum internal activity or traffic within the network.

- Non-substitutable services to the public, i.e. there are few competing networks delivering the same service. (The more effective the existing systems, the more likely its monopoly.)

- The objects of public anxiety about the possible widespread loss of capacity and interrupted service. (The more effective it is, the more likely the anxiety.)

- The source of alarm about the consequences of failures to users and outsiders of serious operating failures, e.g. mid-air collisions, nuclear power station disruptions, etc., and subsequent public expressions of fear and demands for assurances of reliable operations.

3.7

Changing Technologies in Clothing Manufacture

E.A. Rhodes

■ Introduction

The clothing industry operates towards the end-point of a long and complex system of production and distribution which increasingly functions on a global scale (see Article 2.8). Much of the UK industry thus competes with producers across the world, many of whom have the advantage of very low labour costs in what is still, in most areas of production, a labour-intensive industry. It has survived against this competition for three main reasons. First, there are still some advantages in location close to the retail market in a sector which is highly dependent on a short retailing cycle (see section on production diversity, below). Second, concentration in the retail sector has, in some cases, compounded the advantages of proximity to the market – although this also has adverse consequences (see Article 2.8). Third, the high level of diversity in the total product market offers many 'niches' in which firms can specialize. This is particularly evident in the contrast between low cost, mass clothing in basic, standardized designs and the world of haute couture and Saville Row tailoring. But diversity is also evident across the spectrum of types of products which extends from everyday apparel to very highly specialized sportswear, medical and industrial apparel. In the latter categories, the technical specifications of the fabrics used in garment construction and the manufacturing specification can be very demanding, particularly where protection against hazards to health and life are involved.

In contrast to this diversity, the technologies of production were, for a very long time, remarkably homogeneous in character. But this homogeneity has receded during the last 20 years or so as new types of equipment and changes in method have become available. The changing nature of the production process, and its implications for the range of work tasks in garment manufacture, provides the main focus for this article. But first, the underlying character of clothing manufacture needs to be established. This derives from the dynamics of the market pressures outlined in Article 2.8 combined with the processes of clothing design.

■ The design–manufacturing interface

As is to be expected in the context of product diversity, the nature of design tasks in the clothing sector varies widely. Specialist suppliers making, say, protective clothing for car racing, professional, deep-sea divers and the military have to meet, and continually advance, a demanding set of technical requirements. Similarly, leading designers in the area of apparel for sports – such as cycle racing, winter sports, and rock climbing – have sought to exploit developments in manufactured fibres and in fabrics to develop products which will contribute to enhanced participant performance and safety. In the area of 'mainstream' apparel, on which this article concentrates, the tasks are rather different. As emphasized in the earlier article, design in much of this area is fashion-driven. The 'drive' in this respect emanates from a number of sources including some of those depicted in Figure 1. The high fashion of the couture houses is far removed from the average shopping mall and consumer. But, via the fashion and trade press at least, it does have a considerable influence on the

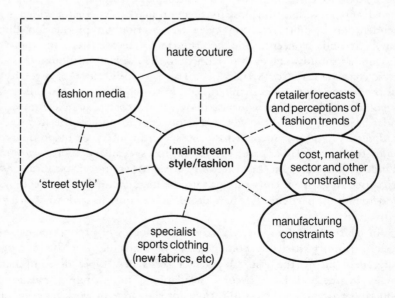

Figure 1   Schema of main influences on mainstream fashion design.

styling choices of 'mainstream' fashion designers and on the more market-oriented concerns of fashion forecasters and buyers. This influence is particularly evident in women's clothing in such respects as preferences for particular garment shapes or silhouettes, and for fabric colours and types (for example, different 'handle', surface texture, weight, print or weave styles) for a particular 'season'.

While popular perceptions of clothing design are style-oriented, in fact, a much wider range of concerns and tasks are involved. Fabric choices and methods of construction have to be related to requirements extending from, say, weather resistance to wearer comfort, and to durability in the cycle of wearing, cleaning and pressing. To meet these requirements involves the determination of fibre and fabric categories, and the selection of processing needs, for instance, to ensure colour fastness and shrink resistance. While many styling changes in the short term may not involve significant changes from one season to another, this task can be substantial, partly because the continuous emergence of new variants in yarn and fibre products offers new styling and other design possibilities. In addition, large retailers have placed emphasis on co-ordination across their product ranges, presenting problems of compatibility in colour, in surface texture and other respects. Thus, despite the apparent simplicity of many garments, design includes a significant element of technical specification. This is particularly important where, as in the example of Marks and Spencer,[1] consistent, high product quality is an instrument in competitive strategy.

Equally important are considerations of manufacturability. Some styles can turn out to be extremely difficult to make in significant volumes at acceptable standards of quality. In the design studio, very highly skilled machinists are employed to produce 'model' or sample garments, generally in a single size. But in the factory, over-complex seam geometries or inappropriate choice of fabric may result in poorly shaped or badly fitting garments when production involves large volumes, and a full range of garment sizes.[2] Product viability in the particular cost and quality area of the market also has to be examined. To achieve this, two particular sets of costs are predominant. The first of these are fabric costs which generally amount to between 40% and 60% of clothing manufacturers' total costs.[3] Correspondingly, both fabric selection and design freedom are constrained by the need to ensure high levels of fabric utilization in manufacture. Similarly, design is constrained by the level of direct labour costs – which may be 30 to 40% of total costs[4] and, in particular, by the concentration of these costs in the assembly area of production where machinists account for about 80% of total labour costs. This distribution creates pressure to minimize the sewing content of designs and, as will be seen below, it has emphasized the application of work study methods and a search for automation.

So, as Figure 1 also seeks to show, what ultimately emerges from the processes of design and decision-making are compromises between fashion designers' inspirations and ideas, the influences of varied external fashion-oriented sources, and the more commercially driven concerns of other parts of the manufacturing and retail organizations. Inevitably, there are tensions around these compromises between what is perceived as creative, and the more cautious, routinized, and 'averaging' concerns of orientations geared to markets and to manufacturing. Thus, in clothing manufacture and retailing, as in other sectors of industry, there is a growing recognition of the penalties associated with 'over the wall' methods, where design decisions are taken in isolation from manufacturing, often resulting in delays and additional costs when the consequent problems are sorted out. In place of the compartmentalization of functions, it is acknowledged that decision-making on design needs to be integrated, i.e. to 'design for manufacture'.

In their totality, design decisions are thus a compromise between the fashion orientation and other types of factor shown in Figure 2. The need to identify what appears to be the best compromise between these varied considerations contributes to a surprisingly long design-to-production cycle in relation to the generally short retail lives of many products (see above, page 163). In total, the production cycle can be a year or more. In part, this reflects the high risk levels associated with clothing design in the fashion and seasonal areas where product redundancy is high. Losses from unsold stock can be considerable. To try and avoid these, a substantial period of time is allocated to design evaluation and revision. Also important is the complexity of the production chain (see Article 2.8), including the comprehensive needs for processing in the various stages. For instance, in the USA, it was estimated that

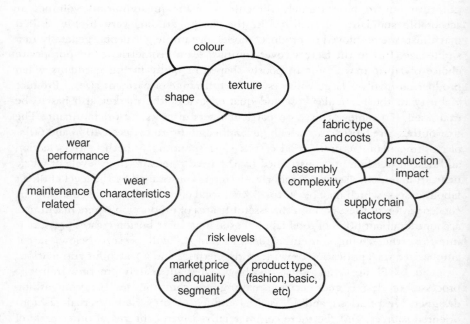

Figure 2   Influences on clothing design.

an average of 65 weeks elapsed between fibre production and customer purchase of finished items.[5] However, it is now recognized that a substantial part of long lead-times results from inefficiencies in the production chain, and firms at all stages are now trying to remove these.

Stages in production

One of the distinguishing features of clothing manufacture is the high level of organizational diversity. In most industries, there are wide disparities in size between the very large plants which may employ several hundred people and 'jobbing' workshops which employ two or three people. These are also evident in clothing, as would be expected from the overall diversity of the product market, ranging from the small-scale, exclusive tailoring enterprise to those concerned with the production of garments in orders of many thousand dozens. But, despite the importance of large volume business, and the presence of some very large firms, there are few very large factories in the UK. Most factories employ less than 150 people, and generally far fewer. In 1985, out of 7000 establishments, only 9% employed more than 100 people, and 75% or more employed fewer than 51 people.[6] Among the smaller enterprises, many operate as sub-contractors to larger firms, undertaking only part of production, primarily garment assembly. Much of this work is undertaken outside the formal economy, sometimes in unsafe conditions, and for very low pay rates, while earnings can fluctuate widely with variations in order levels. While sometimes undertaken in unregistered sweatshops, a substantial part of this type of work is undertaken by homeworkers – an estimated 44% of those involved in women's dress manufacture in the London area are homeworkers.[7]

Homeworking is, of course, almost exclusively confined to women, many of whom are members of ethnic minorities. This reflects the more general employment structure of the industry in two respects. First, it straddles the 'dualist' divide between the primary labour market, where employment security, pay and conditions are relatively reasonable,[8] and the insecurities and poor conditions of the secondary, sub-contracting sector from which even the limited protection of wages council and other legislation was gradually withdrawn during the 1980s. Second, the industry is more generally a women's industry. About 80% of employees are women, most of whom are employed (in skilled work) on the shopfloor – there is a long-standing gender-based division of labour, with males concentrated in maintenance and other technical occupations, and in management.

In the context of diversity in production organization, in product types and markets, there is no 'typical' garment factory – or place of work. However, there are commonalities in the underlying pattern of production. The four basic stages of garment production are shown in Figure 3. Except for the generation of designs and of pattern shapes, these stages will be familiar to anyone who has

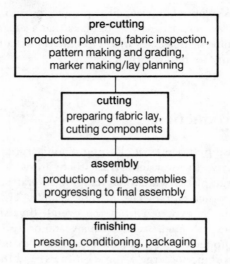

Figure 3 Stages in garment production.

undertaken clothes making in the home, and they are found across the clothing industry, from traditional craft workshops (which survive in bespoke tailoring) to large factories. Until recently, most types of workshop and factory have also used very similar methods, even for very different types of product. The common factor in all of them has been the high level of labour content. This has driven changes in work organization and much of the search for new manufacturing technologies.

The initial stages of work within manufacturing units sit uneasily between design and production, being partly concerned with detail design, modifying designs to maximize production efficiency, but also with generating data for direct use in production. The first stage in this is the creation of pattern shapes for the full size range. These shapes are laid out and manipulated within the fabric contour to produce, by trial and error, a marker (set of templates) for cutting which will maximize utilization of fabric. The fabric used for external and internal components (mainly linings and interlinings) is cut by shears, dies, powered cutting knives or, where volumes are large, by the use of a continuous band-knife. The cut components, together with trims, fastenings and bindings, are then brought together for joining into sub-assemblies which are progressively united into the full assembly. In most cases, joining is by machine stitching. Only in a few cases, mainly involving the insertion of interlinings (which are used to provide support or stability to items such as jacket fronts, shirt collars and cuffs), are other methods of joining used. These mostly involve the use of heat-setting resins and adhesives. Finally, the completed garments are pressed, and packed for transport, storage and display. Where production takes place in factories, as in other industries, a range of co-ordinating, monitoring and support functions operate around the direct production areas, for instance in equipment maintenance, fabric inspection, and production planning.

☐ **Pre-assembly**

As was emphasized above, until recently, nearly all the activities in all four production stages were labour intensive. In the transition from this pattern, the stages before assembly have proved easiest to change. Pattern design and the manual drafting and cutting out of pattern shapes have progressively been moved to CAD systems, first in the larger firms, and then in medium and small firms as system costs fell. For instance, grading and lay planning system costs fell from around £200 000 in the 1970s to between £25 000 and £40 000 in the early 1990s.[9] These systems store on disk the rules applied by a company or a retailer for pattern grading (scaling within the size range) for particular types of designs, and their application enables rapid production of the full range of pattern shapes. In turn, this information is used to simulate the configuration of pattern shapes within a given length and width of fabric. This can be manipulated to search, still by a process of trial and error, for an optimal cutting plan, and has led to significant improvements in levels of fabric utilization. But use of CAD has had different effects – grading has been largely automated and de-skilled whereas in lay planning, the computer serves as an aid, extending the skill and working speed of the operator.

In the following stages, the emerging pattern of practice has diverged. Since the late 1970s, it has been possible to use the data from grading and lay planning to operate automated cutting systems. However, only a limited number of firms had done so by the early 1990s – only 12% of a sample weighted towards the larger and medium-sized firms had installed automated cutters by 1988.[10] This reflects the high cost of these machines which has meant that they have mainly been suitable for large volume production – although some large volume producers felt that there were other disadvantages which militated against purchase. Thus, many firms took an intermediate step, using grading and lay planning systems to drive printers for the rapid production of paper templates, and which contributed to improved performance in conventional cutting systems. However, as key patents expired on the pioneering automated cutters, a new generation of lower cost, more flexible machines has begun to emerge, and it seems probable that the pace of adoption will increase correspondingly.

Overall, the pre-assembly area has moved from being labour intensive to increasing capital intensity. In this process, it has become increasingly linked to the earlier stages of design which have also moved to computer use. Initially, this involved sketching systems which, when allied to scanning devices, enabled the rapid exploration of design concepts. Gradually these systems have extended in type and in capacity. For instance, garment manufacturers can now initiate fabric designs, rather than being dependent on fabric producers' designs; and development is now close to the point where viable 3D design systems will become available. Design systems have also moved towards integration with other systems, for instance to allow rapid estimates of the manufacturing costs of a specific design concept by using data from the firm's materials and production cost databases.[11] The next step is likely to be direct linkage to pattern generation, to the later design stages, and to fabric cutting. The subsequent production stage presents a different set of issues.

□   **Assembly**

As observed above, assembly accounts for a large proportion of non-fabric costs. This, and other factors such as changes in the pattern of demand towards standardized designs, and the shift towards low labour-cost sources of supply (see Article 2.8), have focused attention on cost reduction in work operations in three ways. The first has been an emphasis on raising levels of work intensity. Traditionally, this has been by means of piece-rate payment systems, where operators are paid a specified rate for each dozen work items (the dozen is the general unit for pay measurement and for retail purchasing). But for a long time, the operation of piece-rate systems has been reinforced by the application of work study methods. In part, this has involved work measurement to bring more systematic and incisive assessments of target work speeds and earnings. The use of work study has contributed to a long-standing shift towards increasing task specialization. Thus, in many factories, particularly those concerned with high volume production, assembly work has been subdivided into series of discrete operations, generally concerned with only a limited task of short duration. Work is passed sequentially along the line from initial sub-assemblies to final assembly. One estimate of total cycle times (e.g. picking up a component, positioning it, machining it, and removing it) is that most are of less than 60 seconds' duration.[12] While operators may move between several such tasks, they are likely to work only on one in a single shift. In this, many achieve a frenetic rhythm of work in order to maximize piece earnings but also, by getting ahead of the task norm, to try and gain some 'slack time'.

Despite the pace of work, however, the productivity of operators is perceived as, if not relatively low, capable of substantial improvement. This follows from the modest proportion of work cycle times – about 20% – which are spent on machining parts. Most of the remaining 80% is accounted for by parts handling. Hence, the second type of approach to cost reduction has been to increase the amount of time spent on machining by reducing handling. At a basic level, this has involved workstation design, for instance by identifying the arrangement and size of work tables and other ergonomic considerations which will reduce materials handling time. At a different level, a number of systems for work movement have been developed. The more advanced of these use an overhead transport system to deliver components held in clamps to successive workstations. The parts arrive in positions where they can be rapidly loaded to and from the sewing machine, sometimes without removal from the clamp. The capital costs of these systems is substantial, but they are offset in two main ways. One is an increase in operator productivity – by up to 15% or 20%, depending on the type of product. Second, moving single groups of components (rather than multiples of dozens) and by moving parts to workstations only at the point where the operator is ready to process them, they speed up the flow of materials through the assembly system and greatly reduce the volume of parts (i.e. fabric costs) which are waiting to be processed. As an example of the scale of gains which may be possible in ideal conditions, one equipment manufacturer estimates that throughput times – and thus levels of work in progress – are reduced to 10% of those found in 'traditional' systems using bundles and manual transfer – for instance, from three weeks to three days.

The third type of approach has been to automate production operations. But the nature of textile fabrics presents considerable difficulties in developing systems which are viable. Many of the problems derive from the porosity and limpness of fabrics, which make it difficult to apply the types of robotic and other handling systems that have been utilized in other manufacturing sectors. Robotics applications will depend on sophisticated and costly equipment combinations which include vision sensing and touch-sensitive handling devices. Thus, progress towards the extensive use of mechanical handling systems has been limited, and this has limited moves towards the simultaneous operation of several machines by a single operator (found in many other industries).

A further set of problems is presented by the complex geometry of some garment components and by the continued reliance on stitching.[13] In joining, and in adding stitched elements of design (darts, pleats and so on), most components have to be positioned by hand, and then guided by hand under the sewing head even on the current generations of machines. These and other problems are resolvable, but often only at levels of cost which would put equipment out of reach of all but a very few firms. For example, an American consortium (TC^2) commissioned the development of an automated system for stitching men's coat sleeves and other long-curve seams. Successful prototypes were developed, but the project was abandoned because the high cost and productivity levels of the equipment limited the potential market to only a few firms.[14]

Nonetheless, during the past 20 years or so, there has been considerable progress in developing mechanical auxiliaries, and deploying these in arrays around the sewing head to allow sewing operations to be extended or speeded up. These are important in cost terms down to seemingly minor levels, such as the automatic positioning of labels for insertion into a garment. A number of operations have been mechanized, and in some cases largely automated, primarily where joining involves the stitching of straight seams and in a few cases where combinations of electronic controls and jigs can be applied. For instance, pocket insertion on shirts and jackets, and the stitching of shirt fronts, collars and cuffs can all be undertaken by machine – although, in all these cases, there is a continuing dependence on operators for component positioning prior to the sewing operation. But, in most cases, this equipment is only cost effective for volume production.

Thus, in products such as men's shirts, jeans and trousers, and standard undergarments, manufacturing has become increasingly capital intensive – although with the consequence of a further de-skilling of many operators' jobs. This process has been seen as an emergence of 'islands of automation' in what otherwise remain labour-intensive systems, but which have the capacity to expand into fully automated systems. On the other hand, there are doubts about whether the goal of total automation can be achieved, and about its relevance for the continued production of garments in high labour-cost locations. Capital equipment has become available across the globe with increasing rapidity, and when it is associated with only limited skill requirements, it is difficult to identify sources of competitive advantage in high labour-cost locations.

☐   **Finishing**

The last stage of production involves setting garments in their final shapes, establishing a high standard of finish, final inspection and packaging. This retains a high labour content, but hand ironing and pressing has gradually been displaced in the larger units by a variety of automated and semi-automated pressing and heat treatment facilities.[15] These have been most significant in ensuring consistency in finishing standards and in meeting the increasingly precise requirements associated with 'permanent press' and other fabrics.

■   Towards production diversity?

I have suggested that, until recently, there have been broad similarities in methods and in the types of equipment used which extended right across the diverse spectrum of products and production organizations in the clothing industry. During the past 20 or more years, the development and application of new types of equipment for design, production, and for manufacturing organization and control has opened up a variety of new possibilities. This process offers a number of lessons. One is that, in contrast to early expectations of rapid and dramatic change when new types of equipment and new methods become available, the actual pace of development is generally slow. Hardware and software may be rapidly acquired, but it is often forgotten that these form only one part of a technology. The development of new technologies generally requires the accumulation of much new knowledge, a great deal of which may be specific to the individual firm, and which is acquired by trial and error, sometimes through failures. The obstacles to change are only partly rooted in the technical imperfections of early generations of new equipment and in the difficult processes of learning. Also important is the ability to question the continuing relevance of prior knowledge, experience and assumptions and, where necessary, to recognize and discard that which has no further relevance. But this is a difficult process, and reluctance to abandon established patterns of practice on the part of managers (and others) can present some of the most fundamental obstacles in the path of change.

These types of problem are general, and not confined to the clothing industry. In conceptual terms, they can be viewed in terms of technological paradigms[16] which encompass both a specific body of scientific and technological knowledge, and beliefs about 'best practice' in areas ranging from production methods to shopfloor payment systems. The emergence of radically new technologies presents a challenge to the prevailing paradigm which results in a period of flux and uncertainty in which new technological possibilities are experimented with. The eventual emergence of a new dominant paradigm (incorporating different perceptions of 'best practice') shapes the 'trajectory' of the subsequent search for further developments in the technology.

However, the shift in clothing manufacture appears to be following a different pattern. Developments in hardware and software have opened up a contrasting range of possibilities for the organization of production. Rather than competing for dominance (i.e. as the 'new paradigm'), a number of alternatives have emerged, or are in the process of emerging. Each of these is appropriate to differences in conditions, such as levels of skill within the workforce, labour cost levels, proximity to the main market centres, and product types. An indication of these possibilities is given in Figure 4. From the traditional craft and factory patterns of organization, the main possibilities which have emerged include: the entire relocation of production to low labour cost locations (LCLs); the separation of design and the 'pre-assembly' stages of production from assembly – which is undertaken in either low cost units in the 'home' country[17] or in sites in LCLs.[18] As was described in the section on production stages above, there are also semi-automated units, utilizing dedicated equipment for large volume production. These are sometimes viewed as providing a step towards a further possibility – that of fully automated, but more flexible production systems. However, this concept appears to be some way from realization.

The remaining alternative is that of skill-centred, flexible production systems. In these, CAD intensive design and pre-assembly production are coupled with assembly systems which depend on highly skilled operators, working in teams, and carrying out a wide and varied range of tasks. The emphasis is on low volume production of a mix of products with high levels of added value targeted on the more expensive end of the market. It is this type of approach which has enabled a substantial – though diminished – clothing industry to survive in the high wage conditions of Germany.[19]

Which of these diverse existing and emerging production types seems most likely to be the most appropriate in the UK over the longer term? As was

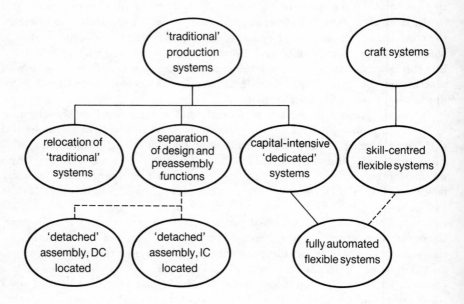

Figure 4   Emerging production alternatives.

indicated above, the answer is partly related to product-specific factors. For much of the UK industry, the importance of sub-contracting for the retail sector and associated production of large volumes has favoured investment in semi-automated, dedicated systems. But, while the European Single Market may appear to offer greater opportunities for standardized products, the continuing relevance of these systems is open to doubt in many product areas. The equipment is widely available, and standardized products are highly suited for the long communications lines that are involved in DC production. Furthermore, retailers, who are also pressed by tight competitive conditions, have been moving towards greater variety in product offerings. At the same time, the more competitive retailers are seeking cost reductions from a switch to 'Quick Response' systems of production and delivery. These involve large reductions in design-to-production lead times (to, say, 16 weeks), and the rapid production of orders in small, repeat batches. Apart from an initial pre-retail order, production takes place in response to the pattern of actual consumer demand, rather than being produced for forecast demand levels prior to retailing. Manufacturers who are close to the home market and who can offer the flexibility to produce mixed orders in comparatively small volumes have an advantage in both these respects. In these circumstances, firms producing middle range, short-life products seem most likely to succeed where they are able to develop a further variant, combining some of the strengths of both 'skill-centred' and dedicated systems depicted in Figure 4. These will provide the flexibility to manufacture mid range, quality garments in small batches, undertaken in a mix of product types.

Notes

(1) Marks & Spencer (M & S) have long employed considerable numbers of technologists with expertise across the textile chain. Their concern extends beyond the existing generation of products to include developments in product and manufacturing technologies (such as new types of fibre and new production equipment and software) to enable M & S to derive early advantage in cost reductions, or improvements in quality or technical performance.

(2) It is important to appreciate that garment sizing does not involve a simple process of sizing up or down from the geometry of the original size.

(3) As 'final assemblers', clothing manufacturers are not unusual in this respect. It appears to be fairly typical for final assemblers to add only around 30–40% of the total value of finished products. This proportion can be expected to continue to fall as production technologies become more complex and specialist suppliers develop new niches within the production chain, and large firms retreat from fully integrated production.

(4) As found, for example, by Hoffman, K. and Rush, H. (1986) *Microelectronics and Clothing: The Impact of Technological Change in a Global Industry*. Science Policy Research Unit, University of Sussex.

(5) Office of Technology Assessment (1987) *The US Textile and Apparel Industry: A Revolution in Progress*. US Department of Commerce.

(6) Carr, H. and Latham, B. (1988) *The Technology of Clothing Manufacture*. BSP Professional Books. Statistics for the numbers of firms and of employees are unreliable, and the figures quoted are an underestimate.

(7) Leigh, R., North, D., Gough, J. and Sweet-Escott, K. (1984) *Monitoring Manufacturing Employment Change in London, 1976–1981*, Vol. 2, Middlesex Polytechnic. Quoted in Phizaclea, A. (1990) *Unpacking the Fashion Industry: Gender, Racism and Class in Production*. Routledge.

(8) I emphasize relatively, because average earnings in the industry are below average, and because employment security tends to be medium term rather than long term – for instance, because of the industry's vulnerability to exchange rate changes and competition from imports.

(9) Cooper, D. and Cooper, M. (1992) Marketing CAD/CAM. In Aldrich, W. *CAD in Clothing and Textiles*. BSP Professional Books.

(10) Whitaker, M., Rush, H. and Haywood, B. (1989) *Technical Change in the British Clothing Industry*. Centre for Business Research, Brighton Business School, Brighton Polytechnic.

(11) Davies, R. (1992) CAD in the real world: using CAD clothing/textiles systems in industry. In Aldrich, *op. cit.*

(12) Coyle, A. (1982) Sex and skill in the organization of the clothing industry. In West, J. (ed.) *Work, Women and the Labour Market*. Routledge.

(13) Stitching has remained the predominant method because the apparent alternative methods of bonding, such as the use of adhesives, have so far proved suitable for only a

limited range of joining. They are unable to meet the requirements of flexibility, appearance and comfort required in garment use, and the rigours of continuous laundering.

(14) However, the project was not regarded as a failure because the development process had generated a number of developments such as control systems which could be utilized to improve other, existing types of equipment on a less ambitious scale.

(15) Whitaker *et al.* (1989) *op. cit.*, found that only a small percentage of firms used automated or semi-automated equipment in finishing, and that most firms had no current plans to introduce this equipment.

(16) This and the related model of technological trajectories (see below) draw on the influential conceptual model developed by Kuhn to explain the processes of development in scientific knowledge (Kuhn, T.S. (1962) *The Structure of Scientific Revolutions*, University of Chicago Press). The relevance of the model for explaining advances in technological knowledge is considered by Dosi in: Dosi, G. (1982) Technological paradigms and technological trajectories. *Research Policy*, 11, 147–62, and in subsequent papers. Dosi defines a technological paradigm as a '. . . "model" and a "pattern" of solution of selected technological problems, based on selected principles derived from natural sciences and on selected material technologies'.

(17) This forms an important part of the production strategy of Benetton – garments are assembled by homeworkers and in small workshops, which are supplied from central design-to-cutting facilities – discussed in Belussi, F. (1987) *Benetton: Information Technology in Production and Distribution: A Case Study of the Innovative Potential of Traditional Sectors*. Science Policy Research Unit, Occasional Paper No. 25, University of Sussex.

(18) For instance, this has already happened in Hong Kong, where some firms act as 'hubs' for assembly undertaken in a number of other South East Asian locations.

(19) Steedman, H. and Wagner, K. (1989) Productivity, machinery and skills: clothing manufacture in Britain and Germany. *National Institute Economic Review*, No. 28, National Institute for Economic and Social Research.

3.8

British Home Stores and its Point-of-Sale System

T102 course team

[One of the points at which most people come into contact with advanced information technology is at the check-out desks in large shops and chain stores. This article describes the development and function of modern 'point of sale' terminals in BHS stores. Each terminal is one of the visible inputs and outputs of the BHS information technology network. Developments in these terminals and networks are taking place very rapidly so that the article is bound to be a little dated when you read it. For example the EPOS and EFTPOS terminals, described in the article, are nowadays in common use in most supermarkets.

(Note that in the article the noun 'store' refers to a 'shop', not to a computer store; though the verb 'storing' refers to the function of a computer memory or store! Also note that 'mainframe computers' are large computers; 'minicomputers' are fairly large computers, and the 'small computers' housed in POS terminals are comparable in power to typical personal PCs or 'micros'.)]

[This article concentrates] on the way British Home Stores (BHS) uses information technology to help run its business.

[. . .]

■ Background to BHS

British Home Stores began as a private company in April 1928. It was launched by American financiers who, having surveyed the British market, decided that there was room for another variety chain store along the lines of the Woolworth operation.

The marketing strategy in the pre-war years was based on establishing a price ceiling and selling a wide and unstructured range of merchandise within that price limit. In 1928 British Home Stores fixed the maximum price of any item at one shilling (5p) thus allowing more flexibility and a selection of lines not directly in competition with Woolworth's whose ceiling was sixpence (2.5p). One year later the ceiling was raised to five shillings (25p) to enable the drapery departments to be introduced.

Central purchasing operated even in those days. All stores carried certain basic lines but each store manager was wholly responsible for ordering his own merchandise and it was up to him to promote the lines that would enable him to achieve his profit target.

The war brought an end to the price ceiling. In a reassessment of strategy, it was decided that price alone would not be the pivot of their marketing methods: it would be replaced by a policy of quality and value for money.

For a long time BHS was seen as the nearest rival to Marks & Spencer (M & S). As William Kay, city editor of *The Times* pointed out in 1985:

'There is no doubt that M & S became an obsession at BHS's ultra-functional head office opposite Marylebone Station.

It was a game BHS could not win. It was always destined to come second to M & S in what was in any case an unfair comparison. But the alleged rivalry also had the advantage of giving BHS a little of the reflected glow from the M & S halo.

The real disadvantage, however, has been that the comparison prevented the top management from thinking independently. M & S was its base line and its yardstick of success.' (Kay, W. (1985) 'How the Conran magic wand could transform BHS', *The Times*, November 26.)

In the early 1980s quite dramatic changes began to alter the appearance of the High Street. New 'design-led' chains such as Next and Principles opened up with immediate success. Existing chains such as the Burton Group embarked upon major face-lifts. Paul Smith, Senior Personnel Executive of Marks & Spencer, describes the reasoning behind these changes from a retailer's point of view:

'Shopping is no longer an activity carried on out of pure necessity. It is a social activity which requires modern retailers to attend to customer needs in terms of merchandise (which was done anyway) and more importantly, to the problems of creating the sort of ambience and environment in which their customers will feel comfortable and hence be more likely to spend.' (Smith, P., 'Retail managers need a flair for making shopping a pleasing social activity today', *The Times*, July 17, p. 28.)

So in 1983 BHS undertook a major research programme in order to find ways of increasing the volume of their sales. Following on from this they too began a major refurbishment programme, spending £60 million in 1985 with a similar spend planned for each of the following four years.

However, the refurbishment plans were, to some extent, soon overtaken by other events. In January 1986, British Home Stores merged with Habitat/Mothercare plc to form Storehouse plc under the Chairmanship of Sir Terence

Conran. Under his guidance the company tried to establish a completely new identity. BHS replaced British Home Stores as the company name. In addition, a different store design was put into all BHS stores in the space of just one weekend in September 1986.

[. . .]

■ Make – distribute – sell

I shall now make use of activity-sequence and flow-block diagrams to describe what happens at BHS.

At the most basic operational level, any retailer must carry out three activities:

(1) obtain goods
(2) display the goods
(3) sell the goods.

The flow of goods will usually be as shown in Figure 1. But if we look at the information that is required in order to carry out these three activities, the flow is in the direction shown in Figure 2. Because this article is concerned with the use of information technology within BHS, I shall start my description of their

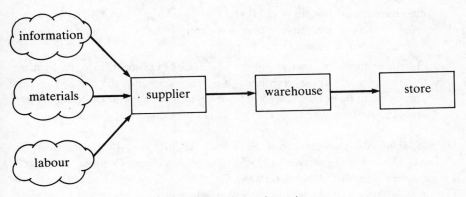

Figure 1   Flow of goods.

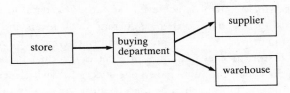

Figure 2   Flow of information.

activities where most of the information is generated – in the store at the point at which payment is handed over by the customers. [. . .]

If you are like me, the only interest you take in the 'tills' when you are out shopping is confined to finding them and estimating which has the shortest queue (and invariably getting it wrong). But the next time you are in the High Street have a look at the tills in a few different types of shops. Examples of the old and the new are shown in Figure 3. [. . .]

Starting with the technology of over twenty years ago and moving forward in time, you are likely to find:

- manual cash registers, using finger-power to operate their mechanics and capable of only simple calculations, if any
- electromechanical cash registers, similar to manual cash registers but driven by electric motors
- electronic cash registers where the mechanical moving parts have been replaced by circuit boards and microchips – these machines allow analysis of sales by type of product, method of payment, volume throughput and promotional codings and undertake calculations for multiple purchases and giving change
- point-of-sale (POS) terminals linked to computers or containing their own microprocessors – after data has been keyed in, these draw on stored information to calculate and process transactions and product information
- electronic point-of-sale (EPOS) terminals, similar to POS terminals but using electronic data capture (magnetic strip or bar-code reader) instead of manual input of data
- electronic funds transfer at point-of-sale (EFTPOS) terminals, replicating the features of EPOS terminals but also allowing the cost of goods purchased to be debited directly from customers' bank or building society accounts and credited to the retailers.

Point-of-sale systems (both POS and EPOS) have been described as 'probably the most significant introduction in retail employment since self-service stores were introduced' (Hines, C., 1978, *The Chips are Down*, Earth Resources Ltd). One of the main aims of introducing such systems is to increase the level of service to the customer whilst at the same time reducing stock levels. Better stock control is achieved by gathering data on sales as the sales take place, using that data to deduce current stock levels and then triggering the replenishment process at the most appropriate time. By providing accurate and reliable historical sales data on a product-by-product basis it also provides information on which products have been successful, which less successful and so on. This information can then be used to guide future buying decisions. One major retailer even claims that the information generated by its EPOS system

Figure 3 Cash registers and terminals.

enables it to improve the quality of its products: 'By allowing decisions to be made earlier it gives people more time to get the products right.'

If you look at Figure 4 you will see the growth that has taken place since then in the number of EPOS systems in the UK. BHS was the first major retailer to introduce POS terminals. Over a two to three year period beginning in 1979 they installed almost 3000 terminals into their 129 stores. During the run-up to the introduction they investigated both POS and EPOS systems. At the time, the only EPOS terminals available were those that used wands to read a magnetic strip. The technology was largely untried (indeed the first EPOS trials only took place between 1975 and 1978 and the first ever on-line application of an EPOS system began in the USA in 1979) and BHS's own experiments led them to believe that manual data entry using a keyboard would, on average, be as fast and as accurate as electronic data capture using a wand.

[. . .]

Imagine you have selected two or three garments, tried them on, liked the look of them, and now you want to pay. At the 'till', or more correctly the POS terminal, the assistant will carry out the sequence of activities shown in Figure 5.

Each POS terminal has a small computer, complete with storage, within itself, but to function fully it has to be connected via a 'local area network' to a 'controller'. BHS has IBM controllers located within its 60 larger stores (see Figure 6). The POS terminals in a store without a controller (a remote store) are linked to the controller in another store (a controller store) by a specialized network using telecommunication lines leased from British Telecom. Each controller is a minicomputer, capable of supporting up to four stores, though in practice it is connected to a maximum of three. The controller in the Milton Keynes store, for example, is connected to 26 POS terminals in its home store

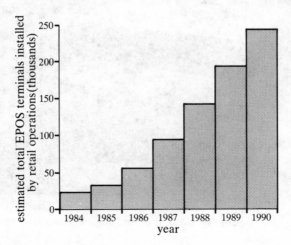

Figure 4   Growth of electronic point-of-sale in the UK.

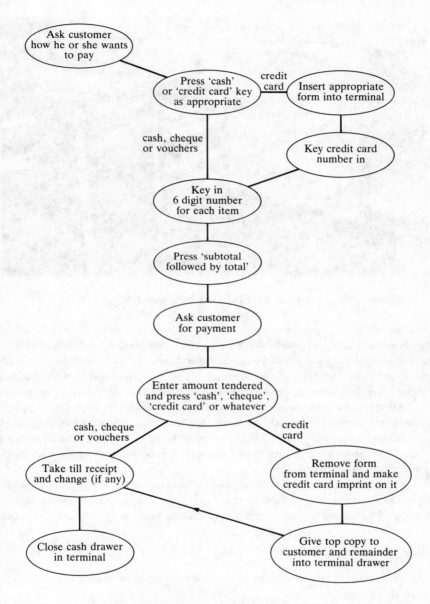

Figure 5 Operation of a POS terminal.

and a further 30 divided between stores in Bedford and Luton. The factor that was most dominant in the decision to limit the number of controllers to 60 was cost. At the time the system was introduced, the price of each controller was far greater than it would be today. So BHS had to weigh up the costs of providing

Figure 6　IBM in-store controller.

one minicomputer per store versus the communications and other costs of sharing computers via British Telecom leased lines. At that time the computers cost more than the telecommunications. The situation today would almost certainly be the reverse, given the fall in the price of computing power.

All stores have the capability of reverting to the public switched telephone network in the event of leased line failure and have procedures for trading in the event of the controller failing. These procedures are, however, fairly laborious to operate, the greatest difficulty being that the items in the shop are not usually individually price-tagged.

Each controller is capable of storing several days' worth of transaction information, various files such as the BHS price file, and some store reports. (We shall be looking at some of these reports and at the way they are used later.)

[. . .]

Now let us look at what happens to the information that is keyed into the POS terminal by the assistant.

If a customer indicates that payment will be made by credit card, the number of the card will automatically be checked against a list of 'negative files' stored in the controller. The negative files are transmitted to each controller daily from the company mainframe before trading begins and usually consist of almost 50 000 card numbers that have been supplied by Access, Storecard, Visa, and so on. The numbers are those of lost or stolen cards and their use should trigger a call to the credit card company concerned.

The six-digit number that is keyed in for each item is almost unique to each line (type or style of goods) stocked. Very similar lines such as differently patterned ties may share a code in order to conserve the stock of codes, which is limited to 99999 because the six-digit code includes a check digit. So if a particular style of skirt is sold in four sizes and three colours, 12 different codes will be needed so that the controller can distinguish between a blue size 12 and a grey size 14, say. When the number has been keyed in it is fed to the controller which 'looks up' the price of the item in its list of about 97 000 price look-up items, and prints it out on the sales receipt, together with an eleven-digit number and a brief description of the item. The eleven-digit code identifies the department from which the item came, its lot number which would enable the item's history to be traced back to manufacture, its colour and its size.

The controller does not keep a record of the items that have been 'looked up' during the day. The codes input into a particular terminal are stored within it until the end of the day. When the store has closed the codes are transferred to the controller as the terminal is 'signed off'. The sales data in each of the 60 controllers is then in turn transferred to BHS's IBM mainframe computer in Luton via an IBM-designed private data network. This same network is then used to transmit price changes and details of new stock to the controllers.

When all the data has reached Luton it is brought together and summarized so that it can be used to prepare two sets of sales reports. One set concentrates on the sales performance of the stores given as daily and cumulative weekly sales by store, department and even individual POS terminal. The other set looks at the level of sales achieved by individual lines and groups of lines.

By 1985/86 BHS's POS system was processing 59.5 million transactions per year, and the company was already conducting experiments into various EPOS and EFTPOS systems with a view to replacing it. Among the factors driving this research are:

- a desire to incorporate new technology – such as automatic cheque printers and credit-card wipers which use wands to read the magnetic strip on cards

- ergonomic and aesthetic reasons – modularization of the terminal (i.e. separating the keypad from the drawer and display), bringing the operator to the level of the customer so that more eye contact can take place between them, making the appearance more in keeping with the design of the store, and so on

- systems/operational reasons – such as enabling the controllers to be programmed in the same language as the mainframe systems, thus giving rise to the ability to offer more sophisticated remote help and diagnostic facilities.

3.9

Radical and Incremental Innovation: A Tale of Three Trains

R. Roy and S. Potter *et al.*

[This article illustrates the dangers of too large a step forward in technological innovation when insufficient resources and time are devoted to the project. It is to be contrasted, for example, with the story of the compact disc (Article 3.3) where a comparable leap forward was successfully taken.]

Back in 1968 British Rail announced plans for a radical new train, the 155 mph Advanced Passenger Train (APT). As is well known, the design of this train represented a radical break from conventional railway engineering [see Figure 1]. It incorporated a host of technical innovations, including body tilt to go round curves at high speed, a new design of brakes and a totally new lightweight construction method. Even the toilets were new and innovative.

Yet it is equally well known that the APT was a failure. It was continually plagued with technical bugs and faded from the public scene in the early 1980s. The incrementally designed 140 mph InterCity 225 replaced the radical APT and was successfully introduced into service in 1989 [see Figure 2].

The APT was born out of a major technical breakthrough in the design of a train's bogies (wheelsets). Conventional bogie and suspension designs had limited the top speed of a train to around 100 mph due to two main factors. Firstly, above this speed the wheelsets develop unstable vibrations and secondly the momentum of fast trains, particularly on curves, causes unacceptable track wear and damage. Using traditional design methods, this limit could be pushed up to 130 mph (but no further) by building an entirely new line with very gentle curves. This is what the Japanese did for their 'bullet' Tokaido line trains, which started running in 1964.

The breakthrough [with the APT] that paved the way for all other fast train projects came in the early 1960s at British Rail's Railway Technical Centre in Derby. By using computer modelling to address the problems of the unstable

Figure 1   Advanced Passenger Train (APT).

Figure 2   InterCity 225.

vibrations and the forces exerted on the track, bogie and suspension systems were devised that raised the potential top speed of trains to 200 mph or more.

In Italy, France and Germany, new stretches of line have been built to exploit fully the potential of this breakthrough. But in Britain the construction of new straight track was politically and economically out of the question. British Rail engineers therefore set about designing a high-tech train that could run as fast as possible on ordinary lines. This required not just the high-speed bogies, but a host of other innovative features – a tilting body, water turbine brakes, ultra-lightweight construction – and those lightweight loos! All these, and more, were built into the 155 mph APT design. By way of contrast, because it would run on new track, France's 162 mph TGV involved a much simpler design.

The project got off to a promising start. But due to wrangling with the government over funding, the timetable slipped, and the APT prototypes did not make their first run until 1979. In that year an APT set the British rail speed record at 162 mph, but it was never reliable enough for regular passenger service. Tilt failures, problems with brakes and multiplicity of bugs due to poor quality control plagued the development team. After one embarrassing attempt at a scheduled service in December 1981, the most the APTs ever saw of passengers was as unadvertised relief trains.

The APT was a technically complex project, but its main problems were to do with managing the project rather than the technology itself. The design and development team became isolated and received little help from a number of traditional railway engineers who viewed the APT project as a threat to their status and working methods. Support for development work was patchy and the quality of construction work often left much to be desired. The final nail in the APT coffin came when the project group was broken up as part of a major departmental reorganization. By the time a new project manager and team had been re-established, the business and design specifications were a decade out of date!

What saved British Rail's InterCity services was that, in addition to the radical APT, an incremental development using the new bogie design was built. This was the 125 mph High Speed Train (HST), or InterCity 125 as it is more widely known. Nearly 100 of these trains became the backbone of four out of the five main InterCity routes in this country. When introduced in 1976 the HST was second only to Japan's Tokaido trains for speed and is still the fastest diesel train in the world.

Beside the use of the innovative high-speed bogie, every other design feature on the HST was nothing more than an incremental improvement on existing 100 mph train designs. Since their initial introduction, the trains have been upgraded and improved while in service. Initially they required a lot of maintenance and there were serious problems with the engines overheating. These problems were overcome and maintenance methods improved. The mileage between power car overhauls has been more than doubled to over 450 000. Each HST now averages 220 000 miles a year. The coach interiors have been completely refurbished, catering provision updated, telephones installed and special accommodation for disabled passengers provided. Coupled with incremental improvements to raise track speeds, the HST today is a considerably improved train over that introduced in 1976. This illustrates the value of continually improving a design to ensure that it retains its market attraction.

To return to the APT; in 1984 the APT was replaced by the InterCity 225 project. A strong management structure was established and an incremental approach adopted in order to get the InterCity 225 into service as quickly as possible. A marketing study suggested that commercially it was not worth exceeding 140 mph as the time savings involved would generate insufficient additional income to pay for the cost of higher speeds. A cut of 15 mph compared with the APT may not seem much, but it allowed the use of a much more conventional design of train, something more akin to the technology of the TGV. Such a train would not have to be as light as the APT and could have conventional brakes. Also, for the first line on which it operates, it would not need to tilt. Tilt is an option which could be developed later. The main innovation is in the locomotive, where, in order to reduce track damage, the weight not cushioned by the suspension system had to be minimized. The train's builders, GEC, developed a new transmission system to do this, which is essentially the only major innovation in this train's design.

The design, development and construction of the 'Class 91' locomotive for the InterCity 225 took only two years (1986–88) and it entered passenger service in 1989. Even the train's introduction has been incremental, first with existing HST coaches and then at 125 mph with new coaches. The full 140 mph package is due to follow in 1993 when track signalling for this speed is complete.

One clear lesson from the APT/HST/InterCity 225 saga is that the gradual introduction of innovations on succeeding versions of a product is much less risky than trying to make the big leap and introduce them all together. A second lesson concerns the technology-led aspect of the APT's development. The APT did not arise because British Rail's passenger planners produced a market specification showing that a 155 mph train was needed. It arose because the engineers offered a 155 mph train. When, ten years later and following the failure of the APT, a proper market specification was developed, it showed that a very different sort of train was really required. The blending of technology-push and market-pull stimuli to innovation needs to be carefully managed so as to avoid the extremes of technically spectacular but unmarketable products being made, or marketing being entirely unaware of what technical progress has made possible. Close and regular contacts between all departments in product development and the use of interdepartmental project teams can aid the creation of a good blend.

The third relates to the concept of lean and robust designs. The APT, and to some extent the HST also, were lean designs – they were intended to fulfil a particular specific role and their basic design concept did not have sufficient flexibility to cope with major variations from that specific role. For example, both the HST and APT are 'fixed formation' trains, consisting of two power cars permanently coupled to a fixed set of coaches. The power cars cannot be used separately from the coaches and it is very difficult to change the number of coaches on a train. In recent years, with overcrowding affecting many HST-operated routes, this inflexibility has been quite a problem.

The InterCity 225, however, consists of a free-standing locomotive, which can be used separately and the locomotive is powerful enough for several more coaches to be added. If required, the design has considerable stretch. It could be upgraded to 150 mph relatively easily, or even faster if two locomotives were coupled to a set of coaches. Overall, although the design of the train has been developed with a particular 1980s view of the InterCity rail market in mind, it is sufficiently robust to adapt to changing circumstances. When markets are liable to unpredictable change, such robust flexibility is valuable.

This robustness is important if the train is to be continually improved in service, as has happened with the HST fleet. Lean designs make it more difficult to keep the process of product improvement going as their 'tight' design closes off a number of options.

Index